(continnued from front flap)

nonlinear, continuous, and discrete time systems as well as stochastic and distributed parameter systems. Many examples and case studies are included.

Daniel Tabak is Associate Professor and Chairman of Automatic Control at Rensselaer Polytechnic Institute of Connecticut, Hartford, Connecticut.

Benjamin C. Kuo is Professor of Electrical Engineering at the University of Illinois, Urbana, Illinois.

PRENTICE-HALL INSTRUMENTATION AND CONTROLS SERIES

CHEN & HAAS *Elements of Control Systems Analysis: Classical and Modern Approaches*
KUO *Analysis and Synthesis of Sampled-Data Control Systems*
KUO *Discrete-Data Control Systems*
OGATA *Modern Control Engineering*
OGATA *State Space Analysis of Control Systems*
TABAK & KUO *Optimal Control by Mathematical Programming*

OPTIMAL CONTROL BY MATHEMATICAL PROGRAMMING

PRENTICE-HALL INTERNATIONAL, INC., *London*
PRENTICE-HALL OF AUSTRALIA, PTY. LTD., *Sydney*
PRENTICE-HALL OF CANADA, LTD., *Toronto*
PRENTICE-HALL OF INDIA PRIVATE LIMITED, *New Delhi*
PRENTICE-HALL OF JAPAN, INC., *Tokyo*

DANIEL TABAK

Associate Professor of Automatic Control
Rensselaer Polytechnic Institute of Connecticut
Hartford

BENJAMIN C. KUO

Professor of Electrical Engineering
University of Illinois
Urbana

OPTIMAL CONTROL BY MATHEMATICAL PROGRAMMING

PRENTICE-HALL, INC.
Englewood Cliffs, New Jersey

© 1971 by Prentice-Hall, Inc., Englewood Cliffs, N. J.

All rights reserved. No part of this book may be reproduced in any form or by any means without permission in writing from the publisher.

Current printing (last digit):
10 9 8 7 6 5 4 3 2 1

13-638106-5

Library of Congress Catalog Card No. 75-137985

Printed in the United States of America

To our wives
Pnina and Margaret

PREFACE

Optimal control is one of the most active research areas of modern technology. It is applied in a variety of fields, such as aerospace, chemical industry, nuclear reactors, transportation and many others. A considerable number of textbooks and monographs covering this subject has appeared during the past ten years. One of the major practical problems in this field is the numerical solution of optimal control problems. Numerous techniques for solving this type of problems have been developed recently.

During the years following World War II a powerful technique for solving finite dimensional optimization problems was developed; namely, mathematical programming. Initially it was applied mainly in operations research problems. Only during the past nine years some effort was done in applying mathematical programming techniques in numerical solutions of optimal control problems. It was demonstrated in many practical cases that a numerical solution, utilizing mathematical programming, was obtained where other methods failed. The material covering the work in applying mathematical programming to optimal control problems is scattered in various journals, theses and reports. It is the purpose of this book to bring this material together under a single cover and a unified presentation. The core of the book originates from a Ph.D. thesis prepared by DT under the advice of BCK at the University of Illinois. This research was later carried on by DT at the Wolf Research and Development Corporation in Riverdale, Maryland. Considerable part of the book is based on the contributions of other workers in this field.

The authors are indebted to the following contributors in the field of mathematical programming applications in optimal control for kindly permitting them to include the results of their work in this book: Professor K. A. Fegley (U. of Pennsylvania), Dr. G. Porcelli (Westinghouse), Professor J. B. Rosen (U. of Wisconsin), Dr. A. F. Fath (Boeing), Lt. Col. C. M. Waespy (USAF), Professor R. R. Mohler (U. of Oklahoma), Professor L. Lasdon (Case Western Reserve U.), Dr. G. S. Jizmagian (A. D. Little, Inc.), Mr. R. W. Harrison (Leeds & Northrup), Dr. A. J. Calise (Raytheon Co.), Professor D. A. Pierre, Dr. V. Lorchirachoonkul (Montana State U.), Professor B. L. Pierson (Iowa State U.), Professor M. Kim (Cornell U.).

The Institute of Electrical and Electronics Engineers, Inc., The Instrument Society of America, The Western Electronic Show and Convention, The Journal of the Franklin Institute, The Pergamon Press, Ltd. (Astronautica Acta) and Taylor & Francis Ltd., (Int. J. of Control) have kindly allowed the authors to reproduce figures and tables which appeared in their publications.

The work described in Sections 8.3 and 8.4 was performed by DT at the WOLF R&D Corp. under NASA contracts NAS 5–9756–112, NAS 5–9756–138. The authors would like to thank Mr. A. J. Rolinski, Mr. G. C. Winston and Mr. N. A. Raumann (NASA, Goddard Space Flight Center) for valuable discussions concerning the NASA Antenna Tracking Systems.

The part of the book written by DT was prepared at the WOLF Research and Development Corporation, Riverdale, Maryland. The support and encouragement of the WOLF management, in particular that of Mr. Arthur R. Dennis and Dr. William T. Wells are highly appreciated. An excellent job of typing of the manuscript was done by Nancy Gebicke, Connie Thomas and Paula Burkhart, and the figures drawing by Mr. C. P. Donohue and Mr. E. Martinez, all of WOLF R&D Corporation. Helpful assistance was given by Mr. R. L. Snodgrass and Mrs. W. M. Walter of WOLF's Publications office.

The authors would like to thank Professor A. P. Sage (Southern Methodist University) and Professor T. J. Higgins (University of Wisconsin), who kindly reviewed the manuscript and contributed valuable suggestions.

Particular thanks and appreciation are due to the authors' wives Pnina Tabak and Margaret Kuo for their patience and understanding.

B.C.K., D.T.

CONTENTS

1. **INTRODUCTION 1**

2. **MATHEMATICAL PROGRAMMING 9**

    2.1 General Definitions 9
    2.2 Formulation of a Mathematical Programming Problem 9
    2.3 Example of a Mathematical Programming Problem 11
    2.4 Classification of Mathematical Programming Problems 12
    2.5 Convexity 15
    2.6 The Kuhn-Tucker Theorem 19
    2.7 Duality 20

3. **NUMERICAL SOLUTION OF MATHEMATICAL PROGRAMMING PROBLEMS 25**

    3.1 Linear Programming 25
    3.2 Quadratic Programming 37
    3.3 Nonlinear Programming 41
    3.4 Unconstrained Minimization Methods 52

4. **OPTIMAL CONTROL AND MATHEMATICAL PROGRAMMING 59**

    4.1 Introduction 59
    4.2 Formulation of the Problem 60
    4.3 Mathematical Programming and Optimal Control 62
    4.4 Computational Considerations 64

5. LINEAR CONTINUOUS-TIME SYSTEMS 70

5.1 Introduction 70
5.2 Formulation of the Problem 71
5.3 A Linear System With Unspecified Sampling Intervals 73
5.4 Fuel-Optimal Problem 77
5.5 Fuel-Optimal Rendezvous Problem 79
5.6 The Time-Optimal Problem 85
5.7 The Generalized Programming Approach 94
5.8 Application in Computing Optimal Controls for a Nuclear Rocket Reactor 103

6. NONLINEAR CONTINUOUS-TIME SYSTEMS 111

6.1 Introduction 111
6.2 Xenon Poisoning Control of Nuclear Reactors 111
6.3 Higher-order Approximation; A Trajectory Optimization Problem 117
6.4 Computer Adjustment of Parameters of Nonlinear Control Systems 120
6.5 The Interior Penalty Method 131
6.6 Solution of Two-Point Boundary-Value Problems 132

7. LINEAR DISCRETE-TIME SYSTEMS 135

7.1 Introduction 135
7.2 Examples of Minimization Problems 136
7.3 Design of a Digital Controller 141
7.4 Discrete-time Controlled Systems with Unknown Sampling Periods 150
7.5 Sampled-Data Systems with Quantized Control 158
7.6 Concluding Remarks 169

8. NONLINEAR DISCRETE-TIME SYSTEMS 172

8.1 Introduction 172
8.2 Example: A Nonlinear Second-Order System 175
8.3 An Antenna Tracking Control System With a Nonlinearity 178
8.4 A Digitally Controlled Nonlinear System 181
8.5 Design of a Digital Controller for a Nonlinear Sampled-Data System 187
8.6 An Iterative Solution 193

9. STOCHASTIC SYSTEMS 197

9.1 Introduction 197
9.2 Combined MP-Monte Carlo Approach 198

9.3 Statistical Design of Sampled-Data Control Systems 200
9.4 Application in Optimal Estimation Problems 208
9.5 Identification 211

10. DISTRIBUTED-PARAMETER SYSTEMS 219

10.1 Introduction 219
10.2 A Heat Conduction System 219
10.3 Multivariable Distributed-Parameter Sampled-Data Systems 222

AUTHOR INDEX 231

SUBJECT INDEX 235

1

INTRODUCTION

During the past ten years a great deal of research activity has been witnessed in the field of optimal control. As a result, numerous papers and a number of books [1–8] have been published on the subject of optimization techniques. The quantitative design of control systems is no longer a trial-and-error effort but rather a precise science involving applied mathematics and high-speed computers. During the early stage of development, control-system studies were characterized by such tools as stability analysis, frequency response, root locus, phase plane, and describing function. These methods, though widely known and of practical use, can be applied only to nonstringent design problems involving single-variable systems, time-invariant systems, and systems with unconstrained variables.

With the advent of the space age, the control engineer is faced with the challenge of designing a great variety of systems having stringent requirements. Common interests have also been found in the design of social systems, nuclear reactor systems, and transportation systems, which require a considerable degree of sophistication in control theory and technology. Some of the more rigid conditions and requirements which characterize these modern systems are multiple input-output, constraints on state and input variables, stochastic systems, unknown or time-varying parameters, large-scale systems, and time-delay systems.

The starting point of modern control theory is the set of state equations which describes the behavior of the dynamic system to be controlled. For a continuous-data system the state equations are a set of first-order differential equations,

$$\frac{d\mathbf{x}(t)}{dt} = \mathbf{f}[\mathbf{x}(t), \mathbf{u}(t)] \tag{1-1}$$

where $\mathbf{x}(t)$ is an $n \times 1$ vector representing the state of the system, $\mathbf{u}(t)$ is an $r \times 1$ input vector, and $\mathbf{f}$ is a vector-valued function. For a system with time delays in the state variables and/or the inputs, the state equations are a set of differential-difference equations, a typical one being

$$\frac{d\mathbf{x}(t)}{dt} = \mathbf{f}[\mathbf{x}(t-T), \mathbf{u}(t-T)] \tag{1-2}$$

where T is the time delay. For a discrete-data system, the state equations are a set of first-order difference equations,

$$\mathbf{x}(k+1) = \mathbf{f}[\mathbf{x}(k), \mathbf{u}(k)] \tag{1-3}$$

where k denotes the kth time instant.

Usually, the optimal control problem is stated as follows: Given

1. The state equations
2. A set of boundary conditions on the state variables at the initial time and at the terminal time
3. A set of constraints on the state variables and control variables

Determine the admissible control $\mathbf{u}(t)$ so that a performance index or cost function is minimized (or maximized).

The known boundary conditions on the state variables are expressed as $\mathbf{x}(t_0)$ and $S[\mathbf{x}(t_f)]$, where t_0 and t_f are the initial and final times, respectively, and S denotes the target set. Although t_0 is always fixed, the final time t_f is not necessarily given.

Because of physical limitations of system components, constraints on state variables and control variables are frequently necessary. For instance, the amplitude constraint on $\mathbf{u}(t)$ may be expressed as $\|\mathbf{u}(t)\| \leqslant M$. Usually, the performance index is written as a scalar quantity,

$$I = G[\mathbf{x}(t_f), t_f] + \int_{t_0}^{t_f} F[\mathbf{x}(t), \mathbf{u}(t), t]\, dt \tag{1-4}$$

where G and F are scalar-valued functions.

The main theoretical approaches to the optimal-control-system design utilize

1. Calculus of variations [9]
2. Maximum (or minimum) principle [1, 2]
3. Dynamic programming [10, 11]

although maximum principle may be regarded as a special application of the method of calculus of variations.

Since the optimal-control problem is one of "extremizing" some performance index I, the classic calculus of variations method developed by mathematicians hundreds of years ago may be used. Depending on the form of the performance index, the optimization by calculus of variations is generally classified under three standard forms as either a Lagrange problem, a Mayer problem, or a Bolza problem. For instance, if the problem is to extremize

$$I_L = \int_{t_0}^{t_f} F[\mathbf{x}(t), \mathbf{u}(t), t] \, dt \qquad (1\text{-}5)$$

where F is assumed to be continuous with respect to $\mathbf{x}$, $\mathbf{u}$, and t, we have a Lagrange problem. If the performance index is

$$I_M = G[\mathbf{x}(t_f), t_f] \qquad (1\text{-}6)$$

the problem is classified as a Mayer problem. Minimum-time and terminal-control problems are well-known examples of the Mayer problem. For instance, if in Eq. (1-6) $G = t_f - t_0$, by minimizing G we have a minimum-time problem or time-optimal problem. However, by properly defining F and G, we can define a given problem as either a Mayer or a Lagrange problem.

Solving the optimal-control problem by calculus of variations leads to the familiar Euler-Lagrange equation which must be solved subject to specified boundary conditions to obtain the optimal trajectory and control. The principal difficulty encountered in attempting a numerical solution of the Euler-Lagrange equation is that the boundary conditions are specified at two end points. This leads to a two-point boundary-value problem (TPBVP) which is difficult to solve analytically except in some very simple cases. Numerical solution of TPBVP's is considreed one of the most difficult problems.

In 1956 Pontryagin and his co-workers, Boltyanskii and Gamkrelidze, developed the maximum principle for continuous-data systems [2]. Basically, the maximum (or minimum) principle provides a set of local necessary conditions for optimality. These necessary conditions are analogous to the necessary conditions that at the minimum of an ordinary smooth function, the first partial derivative must vanish and the second partial derivative must be nonnegative. In the maximum-principle method, variables analogous to the Lagrange multipliers must be introduced. These variables are often called the *costate* or *adjoint-system variables* and are denoted by $p_1(t), p_2(t), \ldots, p_n(t)$. In vector form, $\mathbf{p}(t)$ denotes the costate vector or the adjoint-system variable vector. A scalar-valued function H, which is generally a function of $\mathbf{x}$, $\mathbf{p}$, $\mathbf{u}$, and t, is called the *Hamiltonian function* of the problem and is defined as

$$H(\mathbf{x}, \mathbf{p}, \mathbf{u}, t) = F(\mathbf{x}, \mathbf{u}, t) + \sum_{i=1}^{n} p_i(t) f_i(\mathbf{x}, \mathbf{u}, t) \qquad (1\text{-}7)$$

A concise statement of the maximum principle of Pontryagin for a particular optimal-control problem is given as follows: Given the system described by

$$\frac{d\mathbf{x}(t)}{dt} = \mathbf{f}[\mathbf{x}(t), \mathbf{u}(t), t] \qquad (1\text{-}8)$$

the boundary conditions are
$$\mathbf{x}(t_0) = \mathbf{x}_0$$
and
$$\mathbf{x}(t_f) \in S \quad \text{(target set)}$$

It is assumed that the final time t_f is fixed in this case. We wish to find the admissible control $\mathbf{u}(t)$ which minimizes the performance index

$$I_L = \int_{t_0}^{t_f} F(\mathbf{x}, \mathbf{u}, t)\, dt \qquad (1\text{-}9)$$

For the sake of simplicity, a Lagrange problem is assumed. The control vector may be subject to a constraint on $\mathbf{u}(t)$.

Suppose that $\mathbf{u}^o(t)$ is the optimal control and that $\mathbf{x}^o(t)$ is the optimal trajectory. Then there exists a costate vector $\mathbf{p}^o(t)$ such that

$$\frac{dx_i^o(t)}{dt} = \frac{\partial H}{\partial p_i^o(t)}\bigg|_{\text{on the optimal trajectory}} \qquad (1\text{-}10)$$

$$\frac{dp_i^o(t)}{dt} = -\frac{\partial H}{\partial x_i^o(t)}\bigg|_{\text{on the optimal trajectory}} \qquad (1\text{-}11)$$

where H is the Hamiltonian as defined in Eq. (1-7).

The optimal control $\mathbf{u}^o(t)$ is chosen such that

$$H(\mathbf{x}^o, \mathbf{u}^o, \mathbf{p}^o, t) \geqslant H(\mathbf{x}^o, \mathbf{u}, \mathbf{p}^o, t) \qquad (1\text{-}12)$$

In addition, a transversality condition of $\mathbf{p}^o(t_f)$ transversal to the target set S at $\mathbf{x}^o(t_f)$ must be satisfied. In other words, $\mathbf{p}^o(t_f)$ is orthogonal to the tangent plane of S at $\mathbf{x}^o(t_f)$. For a more general formulation of the maximum (or minimum) principle for various other cases, the reader is referred to reference 1. A more comprehensive formulation which includes state- and control-constrained problems can be found in reference 6.

Considering only the formulation of a particular optimal-control problem as given above, we can see that the initial and transversality conditions lead again to a TPBVP. The formulation becomes even more complicated in state-inequality-constrained problems [6].

The dynamic-programming method was first developed by Bellman [10] for discrete-data systems. The foundation of the method is the *principle of optimality*, which states:

> An optimal control strategy has the property that, whatever the initial state and the initial decision, the remaining decision must form an optimal control strategy with respect to the state resulting from the first decision.

In the continuous-data system form, dynamic programming is often referred to as the Hamilton-Jacobi-Bellman theory. However, the major advantage of the dynamic-programming method is its discrete form, which makes it convenient for numerical solution. For instance, if we wish to minimize the performance index

$$I_L = \int_{t_0}^{t_f} F(\mathbf{x}, \mathbf{u}, t)\, dt \tag{1-13}$$

we let

$$S[\mathbf{x}(t_f), t_f] = \min_{\mathbf{u} \in \Omega} I_L \tag{1-14}$$

If we let $t = t_0 + i\Delta$, with $i = 0, 1, 2, \ldots, N$, so that

$$t_f - t_0 = N\Delta \tag{1-15}$$

Eq. (1-14) can be written as

$$S[\mathbf{x}(t_f), t_f] = S_N[\mathbf{x}(t_f), t_f] = \min_{\mathbf{u} \in \Omega} \int_{t_0}^{t_f} F(\mathbf{x}, \mathbf{u}, t)\, dt \tag{1-16}$$

Applying the principle of optimality, Eq. (1-16) becomes

$$S_N[\mathbf{x}(t_f), t_f] = \min_{\mathbf{u} \in \Omega} \left\{ \int_{t_0}^{t_f - \Delta} F(\mathbf{x}, \mathbf{u}, t)\, dt + S_1[\mathbf{x}(t_f - \Delta), t_f - \Delta] \right\} \tag{1-17}$$

where

$$\begin{aligned} S_1[\mathbf{x}(t_f - \Delta), t_f - \Delta] &= \min_{\mathbf{u}(t_f - \Delta)} \int_{t_f - \Delta}^{t_f} F[\mathbf{x}(t_f - \Delta), \mathbf{u}(t_f - \Delta), t_f - \Delta]\, dt \\ &\simeq \min_{\mathbf{u}(t_f - \Delta)} \{\Delta F[\mathbf{x}(t_f - \Delta), \mathbf{u}(t_f - \Delta), t_f - \Delta]\} \end{aligned} \tag{1-18}$$

The state equation $\dot{\mathbf{x}} = \mathbf{f}(\mathbf{x}, \mathbf{u}, t)$ may be approximated by the discrete model

$$\frac{\mathbf{x}[(k+1)\Delta] - \mathbf{x}(k\Delta)}{\Delta} = \mathbf{f}[\mathbf{x}(k\Delta), \mathbf{u}(k\Delta), k\Delta] \tag{1-19}$$

In general, for the discretized N-stage process,

$$\begin{aligned} S_{N-j}&\{\mathbf{x}[t_f - (N-j)\Delta], j\Delta\} \\ &= \min_{\mathbf{u}} \big(\Delta F\{\mathbf{x}[t_f - (N-j)\Delta], \mathbf{u}[t_f - (N-j)\Delta], j\Delta\} \\ &\quad + S_{N-j+1}\{\mathbf{x}[t_f - (N-j-1)\Delta], \mathbf{u}[t_f - (N-j-1)\Delta], \\ &\qquad\qquad t_f - (N-j-1)\Delta\}\big) \qquad j = 0, 1, 2, \ldots, N \end{aligned} \tag{1-20}$$

where $S_0 = 0$.

Equation (1-20) represents a set of recursive relations which can be solved step by step numerically by a digital computer.

A major disadvantage of dynamic programming is the excessive computer-memory requirement [10, 11]. For instance, if the state vector $\mathbf{x}$ is n-dimensional and each state variable is tabulated by M discrete values over its admissible range, we need to maintain a table of M^n values of S_{N-j} from step to step. In a four-dimensional problem with $M = 100$, we need a table of 10^8 entries. Having more than one control variable makes the search for the optimal value a much more difficult task. Many attempts have been made to reduce the storage requirements of the dynamic-programming algorithm.

One of the most significant programming achievements in this area is the state-increment algorithm proposed by Larson [11]. Considerable reduction in storage requirements is achieved as a result of applying the state-increment method. However, the problem is solved up to a certain point. Optimal-control problems for systems of an order higher than 4 and with multiple control variables are still very difficult to solve by dynamic/programming and in many cases are completely untreatable.

A major effort to find efficient solutions to TPBVP's has been exerted in the past ten years. Only a small sample of the existing literature is quoted here [6, chaps. 13–15; 12–18]. Many classes of problems have been solved by the authors cited and by many others. However, there still exist some major difficulties in solving TPBVP's. In particular, the following classes of problems present major difficulties in establishing efficient TPBVP algorithms:

1. State-control, nonlinearly constrained problems, where constraints of the following type are imposed:

$$g(\mathbf{x}, \mathbf{u}) \leqslant 0 \qquad (1\text{-}21)$$

where $\mathbf{g}$ is a nonlinear vector function.
2. Many classes of nonlinear-multivariable systems pose serious problems in TPBVP solutions even in unconstrained cases.
3. Problems involving discrete-time systems with constraints imposed on the sampling intervals.
4. Discrete-time systems, nonuniformly sampled, where the sampling intervals are unknown a priori.

This list is far from exhaustive.

An approach somewhat different from the ones discussed above is presented in this book. This approach utilizes the techniques of mathematical programming, whose basic notions and numerical algorithms are described in Chapters 2 and 3. A mathematical-programming problem consists of extremizing a multivariable function subject to multiple inequality and equality constraints. The theory and numerical algorithms of mathematical programming have been under continued research and development since the 1940's. These methods were applied mainly in the operations research area. Interest in applying mathematical-programming techniques to numerical solutions of optimal-control problems arose only in the 1960's. One of the earliest works on this subject was done by Zadeh and Whalen in 1962 [19]. Since then, a considerable number of publications on this subject have appeared, most of them during 1967–1969. The details of these works are discussed in Chapters 4–10. Basically, most of the approaches reformulate the optimal-control problem in a form typical of a mathematical-programming problem. As a next step, the problem is solved numerically, utilizing one of the available algorithms, which are described in Chapter 3. In some cases, new algorithms, suited for the particular problems treated, have been generated.

Among the main advantages of using the mathematical-programming approach are

1. Most of the mathematical-programming algorithms handle inequality constraints in an efficient manner. Many complicated inequality constraints, involving both state and control variables, or sampling intervals, which constitute a formidable TPBVP, can be treated with relative ease by a mathematical-programming algorithm.
2. For problems with comparable numbers of variables, the mathematical-programming algorithms do not have as excessive computer-storage requirements as dynamic programming.

It is not implied that the mathematical-programming approach is a universal remedy for numerical solutions of complicated optimal-control problems. Many classes of mathematical-programming problems (particularly nonlinear, nonconvex ones) are very difficult to handle computationally. There exist classes of optimal-control problems where other approaches have turned out to be more efficient. However, for a large class of optimal-control problems, the mathematical-programming approach is, to date, the most efficient one—and in some cases the only one actually worked out in practice.

The application of mathematical-programming algorithms in various classes of optimal-control problems is presented in Chapters 5–10. Chapters 2 and 3 serve as a general introduction to the basic theory and computational methods of mathematical programming. The connection between optimal control and mathematical programming as well as some general computational considerations are discussed in Chapter 4. The following classes of optimal-control problems are treated separately: linear continuous-time systems in Chapter 5, nonlinear continuous-time systems in Chapter 6, linear discrete-time systems in Chapter 7, nonlinear discrete-time systems in Chapter 8, stochastic systems in Chapter 9, and distributed-parameter systems in Chapter 10. In all these chapters (5–10), the discussion centers on the method of applying mathematical-programming techniques to the particular classes of systems and on the presentation of various case studies as examples.

REFERENCES

1. M. Athans and P. L. Falb, *Optimal Control: An Introduction to the Theory and its Applications*, McGraw-Hill, New York, 1966.
2. L. S. Pontryagin, V. G. Boltyanskii, R. V. Gamkrelidze, and E. F. Mishchenko, *The Mathematical Theory of Optimal Processes*, Wiley, New York, 1962.
3. E. B. Lee and L. Markus, *Foundations of Optimal Control Theory*, Wiley, New York, 1967.
4. D. J. Wilde and C. S. Beightler, *Foundations of Optimization*, Prentice-Hall, Englewood Cliffs, N.J., 1967.

5. V. W. Eveleigh, *Adaptive Control and Optimization Techniques*, McGraw-Hill, New York, 1967.
6. A. P. Sage, *Optimum Systems Control*, Prentice-Hall, Englewood Cliffs, N.J., 1968.
7. G. Leitmann, *Optimization Techniques*, Academic Press, New York, 1962.
8. D. G. Luenberger, *Optimization by Vector Space Methods*, Wiley, New York, 1969.
9. I. M. Gelfand and S. V. Fomin, *Calculus of Variations*, Prentice-Hall, Englewood Cliffs, N.J., 1963.
10. R. Bellman, *Dynamic Programming*, Princeton University Press, Princeton, N.J., 1957.
11. R. E. Larson, *State Increment Dynamic Programming*, American Elsevier, New York, 1968.
12. A. E. Bryson and W. F. Denham, "A Steepest Ascent Method for Solving Optimum Programming Problems," *J. Appl. Mech.*, **29**, pp. 247–257, 1962.
13. W. F. Denham and A. E. Bryson, "Optimal Programming Problems With Inequality Constraints II: Solution by Steepest-Ascent," *AIAA J.*, **2**, pp. 25–34, 1964.
14. P. Kenneth and R. McGill, "Two-Point Boundary-Value-Problem Techniques," *Advan. Control Systems*, **3**, pp. 69–109, 1966.
15. D. Isaacs, "Algorithms for Sequential Optimization of Control Systems," *Advan. Control Systems*, **4**, pp. 1–71, 1966.
16. R. E. Kopp and H. G. Moyer, "Trajectory Optimization Techniques," *Advan. Control Systems*, **4**, pp. 103–155, 1966.
17. D. K. Scharmack, "An Initial Value Method for Trajectory Optimization Problems," *Advan. Control Systems*, **5**, pp. 51–132, 1967.
18. J. B. Plant, *Some Iterative Solutions in Optimal Control*, M.I.T. Press, Cambridge, Mass., 1968.
19. L. A. Zadeh and B. H. Whalen, "On Optimal Control and Linear Programming," *IRE Trans. Automatic Control*, **AC-7**, pp. 45–46, 1962.

2

MATHEMATICAL PROGRAMMING

2.1 GENERAL DEFINITIONS

Mathematical programming (MP) [1-6] involves the solution of a multi-variable extremization problem. The variables considered are expressed in a certain functional relationship which has to be minimized or maximized. This expression is usually called an *objective function* or a *performance index*.

In addition to extremizing the objective function, the variables considered are usually required to satisfy additional inequalities or equalities, denoted as *constraints*. Both the objective function and the constraints may in general be nonlinear functions of all or some of the variables considered. In case no constraints are imposed, one has a so-called *unconstrained extremization* problem.

Any set of variables under consideration which satisfies *all* of the constraints of the problem is called a *feasible solution*. A feasible solution which extremizes (minimizes or maximizes) the objective function constitutes an *optimal solution* to the MP problem.

2.2 FORMULATION OF A MATHEMATICAL-PROGRAMMING PROBLEM

Suppose the MP problem under consideration involves n variables x_i ($i = 1, 2, \ldots, n$). The n variables form an n-dimensional *variables vector* $\mathbf{x}$. The objective function of the n variables x_i will be denoted $f(\mathbf{x})$. $f(\mathbf{x})$ is a scalar, generally nonlinear function of all or, in some cases, some of the x_i ($i = 1, \ldots, n$) variables.

The inequality constraints may be formulated as follows:

$$g_i(\mathbf{x}) \geqslant 0, \qquad i = 1, \ldots, q \tag{2-1}$$

where the $g_i(\mathbf{x})$'s ($i = 1, \ldots, q$) are scalar, nonlinear functions of some or all of the x_i ($i = 1, \ldots, n$) variables. The $g_i(\mathbf{x})$'s can be regarded as components of a vector function $\mathbf{g}(\mathbf{x})$, so that (2-1) may be written as

$$\mathbf{g}(\mathbf{x}) \geqslant \mathbf{0} \tag{2-2}$$

Similarly, the equality constraints may be denoted in scalar form:

$$h_i(\mathbf{x}) = 0, \qquad i = 1, \ldots, p \tag{2-3}$$

where the $h_i(\mathbf{x})$'s ($i = 1, \ldots, p$) are scalar, nonlinear functions of some or all of the x_i ($i = 1, \ldots, n$) variables, or in vector form,

$$\mathbf{h}(\mathbf{x}) = \mathbf{0} \tag{2-4}$$

In Eq. (2-2) $\mathbf{0}$ represents a q-dimensional zero vector, and in Eq. (2-4) a p-dimensional zero vector.

The mathematical-programming problem may now be stated as follows:

$$\begin{aligned}&\text{Minimize (or maximize)} &&f(\mathbf{x}) \\ &\text{Subject to} &&\mathbf{g}(\mathbf{x}) \geqslant \mathbf{0} \\ &&&\mathbf{h}(\mathbf{x}) = \mathbf{0}\end{aligned} \tag{2-5}$$

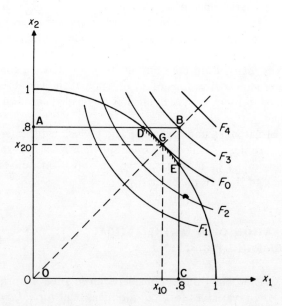

FIG. 2-1. Example of a mathematical programming problem.

or in symbolic form,

$$\min \{f(\mathbf{x}) \mid g_i(\mathbf{x}) \geq 0, i = 1, \ldots, q; h_j(\mathbf{x}) = 0, j = 1, \ldots, p\} \quad (2\text{-}6)$$

The constraints in Eq. (2-6) are occasionally expressed in scalar form so that their number can be seen. Using minimization instead of extremization, one does not make the problem less general. Suppose, for example, one wants to *maximize* a function $f_1(\mathbf{x})$. This is equivalent to *minimizing* a function:

$$f(\mathbf{x}) = -f_1(\mathbf{x})$$

There is also no loss of generality in using inequality constraints of the type in Eq. (2-1). For example, suppose one had constraints in the form

$$g_{1i}(\mathbf{x}) \leq 0$$

Defining a new set of functions,

$$g_i = -g_{1i}$$

one would again obtain the set of constraints in Eq. (2-1).

2.3 EXAMPLE OF A MATHEMATICAL-PROGRAMMING PROBLEM

As an illustrative example of a mathematical-programming problem, a two-dimensional problem with four inequality constraints and one equality constraint is presented. The problem is as follows:

$$\text{Maximize} \quad f(x_1, x_2) = x_1 x_2 \quad (2\text{-}7)$$
$$\text{Subject to} \quad .8 - x_1 \geq 0 \quad (2\text{-}8)$$
$$.8 - x_2 \geq 0 \quad (2\text{-}9)$$
$$x_1 \geq 0 \quad (2\text{-}10)$$
$$x_2 \geq 0 \quad (2\text{-}11)$$
$$x_1^2 + x_2^2 - 1 = 0 \quad (2\text{-}12)$$

The problem can best be explained using a graphical representation in the x_1-x_2 plane (Fig. 2-1). The constraints in Eqs. (2-10) and (2-11) restrict the feasible region to the first quadrant. The constraints in Eqs. (2-8) and (2-9) confine the region within the square $ABCO$.

The equality constraint of Eq. (2-12) implies that the feasible region should be on the unit circle, with the center at the origin. Therefore, the actual feasible region of the problem consists of the segment DE of the unit circle whose equation is given by (2-12).

The curves F_1 to F_4 are loci of constant $f(x_1, x_2)$, so that $F_1 < F_2 < \cdots < F_4$. It is easy to see that these curves are hyperbolas described by

$$x_2 = \frac{F_i}{x_1}$$

and are symmetrical about the line OB. Let the curve which is tangent to the arc DE at point G be designated as F_0. (G is also the crossing point between DE and OB.) Then, at the point $G(x_{10}, x_{20})$, $f(x_{10}, x_{20}) = F_0$ attains its maximal value, while satisfying all the constraints of Eqs. (2-8)–(2-12). This makes

$$x_{10} = x_{20} = .707$$

which is the optimal solution for the posed problem.

The optimal value of the objective function is

$$F_0 = x_{10} x_{20} = (.707)^2 = .5$$

2.4 CLASSIFICATION OF MATHEMATICAL-PROGRAMMING PROBLEMS

Consider the MP' problem as formulated in Eq. (2-6). If $f(\mathbf{x})$, $g_i(\mathbf{x})$ ($i = 1, \ldots, q$), and $h_j(\mathbf{x})$ ($j = 1, \ldots, p$) are linear functions of x_i ($i = 1, \ldots, n$), one has the so-called *linear-programming* (LP) problem [7–11]. It should be stressed, however, that for the problem to qualify as an LP problem, *all* the constraints as well as the objective function must be linear functions of x_i.

In a linear-programming problem the objective function may be expressed as

$$f(\mathbf{x}) = \sum_{i=1}^{n} c_i x_i = \mathbf{c}^T \mathbf{x} \qquad (2\text{-}13)$$

where $\mathbf{c}^T$ is the matrix transpose of the n-dimensional constant vector $\mathbf{c}$. In general, the constraints can be expressed in the following form:

$$A\mathbf{x} \begin{Bmatrix} \geqslant \\ = \\ \leqslant \end{Bmatrix} \mathbf{b} \qquad (2\text{-}14)$$

where $\mathbf{b}$ is an m-dimensional constant vector, m is the total number of constraints, and A is a constant $m \times n$ matrix. A more detailed discussion of the formulation of the constraints in an LP problem will be given in Chapter 3. Usually, constraints involving a restriction on the sign of certain variables, such as

$$x_i \geqslant 0, \qquad i = 1, \ldots, n$$

are written out separately and are not included in the matrix A.

As an example, consider the following LP problem (Fig. 2-2):

Sec. 2.4   Classification of Mathematical-Programming Problems   13

$$\text{Maximize} \quad (x_1 + 4x_2) \tag{2-15}$$
$$\text{Subject to} \quad x_1 \leqslant 10$$
$$x_2 \leqslant 10 \tag{2-16}$$
$$x_1 + x_2 \leqslant 14$$
$$x_1, x_2 \geqslant 0 \tag{2-17}$$

As an alternative, one can write

$$\max \{\mathbf{c}^T\mathbf{x} \mid A\mathbf{x} \leqslant \mathbf{b}; \; x_1, x_2 \geqslant 0\} \tag{2-18}$$

where

$$\mathbf{c}^T = [1 \quad 4]$$

$$A = \begin{bmatrix} 1 & 0 \\ 0 & 1 \\ 1 & 1 \end{bmatrix} \quad \mathbf{b} = \begin{bmatrix} 10 \\ 10 \\ 14 \end{bmatrix}$$

By inspecting the constraints in Eqs. (2-16) end (2-17), one can see that the feasible region is the inside of the polygon $ABCDO$ as shown in Fig. 2-2. The lines L_1, L_2, and L_3 represent the loci of constant values of the objective function (2-15) in an ascending order. It is easy to see that line L_2, for which $\mathbf{c}^T\mathbf{x} = 44$, is the maximal value of the objective function, which

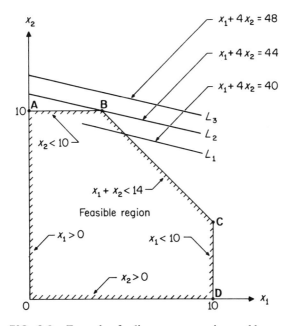

FIG. 2-2. Example of a linear-programming problem.

has at least one common point (B) with the feasible region; i.e., point B is the optimal point. Hence, the optimal solution for this problem is

$$x_{10} = 4$$

and

$$x_{20} = 10$$

A detailed discussion concerning the numerical solution to LP problems will be given in Chapter 3.

If the objective function of a mathematical-programming problem is *quadratic*, whereas *all* the constraints are linear, one has the so-called *quadratic-programming* (QP) problem [1–6, 12]. The objective function of a QP problem can in general be expressed as

$$f(\mathbf{x}) = \mathbf{c}^T\mathbf{x} + \mathbf{x}^T D \mathbf{x} \tag{2-19}$$

where D is an $n \times n$ constant symmetrical matrix. Other variables are the same as in the formulation of the LP problem. The linear constraints of the QP problem may be expressed in the same general way as those of the LP one. The objective function can also be written in the scalar form,

$$f(x_1, \ldots, x_n) = \sum_{i=1}^{n} c_i x_i + \sum_{i,j=1}^{n} x_i d_{ij} x_j \tag{2-20}$$

where d_{ij} denotes the components of the D matrix.

Inspecting Eq. (2-20) and noting that $x_i x_j = x_j x_i$, one may see that no generality is lost by assuming that D is a symmetrical matrix.

The example presented in Sec. 2.3 can serve as an illustration of a QP problem, with the exclusion of the equality constraint (2-12). However, now the feasible region is the inside of the square *ABCO* (see Fig. 2-1), and the optimal point is B ($x_{10} = x_{20} = .8$). The optimal value of the objective function is

$$f(x_{10}, x_{20}) = x_{10} x_{20} = (.8)^2 = .64$$

Some numerical methods for the solution of QP problems will be described in Chapter 3.

The most general case of the MP problem is the *nonlinear-programming* (NLP) problem [1–6]. To belong to this category, it is sufficient for a problem to have at least one nonlinear constraint or to have a nonlinear objective function. Naturally, QP is a particular case of nonlinear programming; however, owing to the linearity of all its constraints, it is considered as a class by itself. The reason for this will become clear during the discussion of computational methods in Chapter 3. The example presented in Sec. 2.3 is obviously an example of an NLP problem.

A particular class of NLP problems constitutes the so-called *geometric-programming* (GP) problem [13, 14]. The general form of its objective function and *inequality* constraints is as follows:

$$\sum_{i}^{m} c_i \prod_{j}^{n} x_j^{a_{ij}}$$

where c_i = constant positive coefficients, $i = 1, \ldots, m$
a_{ij} = arbitrary real numbers, $j = 1, \ldots, n$

The main feature of geometric programming is that the optimal cost associated with each term $c_i \prod_j x_j^{a_{ij}}$ of the objective function may be established separately, before solving for the optimal x_j ($j = 1, \ldots, n$) variables [13, 14].

In some applications involving a solution of a linear-programming problem, the variables are restricted to being discrete integers. This kind of problem is denoted as *integer linear programming* (ILP) [2, chap. 8; 15]. So far, most of the effort put forth with this class of problems has been restricted to the linear problems; hence some authors omit the term *linear* and use only *integer programming* [15]. In some problems, only some of the variables are restricted to being integers; the rest may change continuously. Problems of this type are called *mixed-integer-programming* (MIP) problems [16].

There are problems in which the parameters forming the coefficients in the constraints or the objective function are random variables. These are described as *stochastic-programming* (SP) proplems [1, chap. X; 2, chap. 5; 7, chap. 25].

2.5 CONVEXITY

Convexity [17] is one of the most important mathematical notions used in mathematical programming. Before presenting some basic methods of mathematical programming, the subject of convexity will be discussed, and some fundamental definitions will be given.

As mentioned in previous sections, every MP problem has its own feasible region in the state space, which consists of all of the points satisfying all of the constraints of the problem. It is actually a set of points representing feasible solutions of the problem. Of particular interest in MP problems are the *convex sets*.

Definition of convexity. A set is said to be *convex* if for any two points belonging to the set, the line segment joining them is contained in the set.

Consider the set of points S (Fig. 2-3) and two points A and B belonging to it, i.e., $A \in S$ and $B \in S$. Points A and B represent the vertices of the vectors $\mathbf{x}_1$ and $\mathbf{x}_2$, respectively. The segment AB joining the two points is actually the vector $\mathbf{x}_1 - \mathbf{x}_2$. For any scalar λ such that $0 \leqslant \lambda \leqslant 1$, the vector

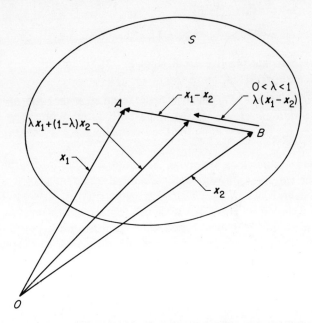

FIG. 2-3. Definition of convexity.

$\lambda(\mathbf{x}_1 - \mathbf{x}_2)$ will be collinear with the vector $\mathbf{x}_1 - \mathbf{x}_2$, as shown in Fig. 2-3. It is easy to see that point P on the segment AB is represented by the vector $\lambda \mathbf{x}_1 + (1 - \lambda)\mathbf{x}_2$. One may now modify the definition of a convex set as follows: A point set S is called *convex* if for any two points $\mathbf{x}_1, \mathbf{x}_2 \in S$, *all points*

$$[\lambda \mathbf{x}_1 + (1 - \lambda)\mathbf{x}_2] \in S, \qquad 0 \leqslant \lambda \leqslant 1$$

In Fig. 2-4(a) are some examples of convex sets and in Fig. 2-4(b) are some nonconvex sets.

Similarly, one may define the notion of a *convex function*. A function $f(\mathbf{x})$ is *convex* over a convex set S, if for any two points $\mathbf{x}_1, \mathbf{x}_2 \in S$, and for all scalars λ, $0 \leqslant \lambda \leqslant 1$ (Fig. 2-5):

$$f[\lambda \mathbf{x}_1 + (1 - \lambda)\mathbf{x}_2] \leqslant \lambda f(\mathbf{x}_1) + (1 - \lambda)f(\mathbf{x}_2) \qquad (2\text{-}21)$$

If the inequality in (2-21) is strict (i.e., $<$), the function $f(\mathbf{x})$ is *strictly convex*. If the inequality sign in (2-21) is reversed, $f(\mathbf{x})$ is *concave*. If the reversed sign is strict (i.e., $>$), $f(\mathbf{x})$ is *strictly concave*. From the definitions one may see that if $f(\mathbf{x})$ is convex (or strictly convex), then $-f(\mathbf{x})$ is concave (or strictly concave), and vice versa.

The importance of convexity in MP problems will become more apparent in later discussions, particularly in Chapter 3. One of the most important properties of convex functions may be seen from the following discussion.

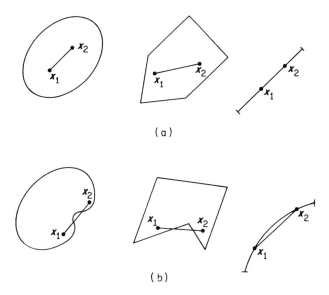

FIG. 2-4. Examples of convex and nonconvex sets: (a) convex sets; (b) nonconvex sets.

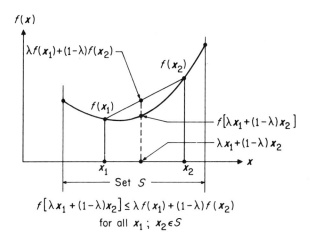

FIG. 2-5. A convex function.

Consider the functions $f_1(\mathbf{x})$ and $f_2(\mathbf{x})$ presented in Fig. 2-6(a) and (b), respectively. The function $f_1(\mathbf{x})$ is convex and $f_2(\mathbf{x})$ is not. Consider both functions in the domain $\{S \mid \mathbf{x}_1 \leqslant \mathbf{x} \leqslant \mathbf{x}_2\}$. $f_1(\mathbf{x})$ has a minimum $f(\mathbf{x}^\circ)$ at $\mathbf{x}^\circ \in S$ such that

$$f_1(\mathbf{x}^\circ) \leqslant f_1(\mathbf{x}) \qquad \text{for all } x \in S$$

In other words, the convex function $f_1(\mathbf{x})$ has a *global* minimum in the domain S.

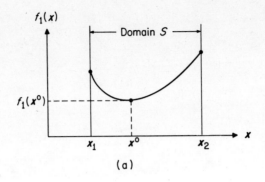

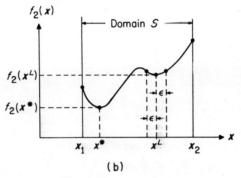

FIG. 2-6. Minimum of a convex and a nonconvex function: (a) convex function $f_1(\mathbf{x})$; (b) nonconvex function $f_2(\mathbf{x})$.

Similarly, the nonconvex function $f_2(\mathbf{x})$ has a global minimum at the point $\mathbf{x}^* \in S$ such that

$$f_2(\mathbf{x}^*) \leqslant f_2(\mathbf{x}) \qquad \text{for all } \mathbf{x} \in S$$

On the other hand, in the neighborhood of the point $\mathbf{x}^L \in S$ the following condition is satisfied:

$$f_2(\mathbf{x}^L) \leqslant f_2(\mathbf{x}^L \pm \epsilon), \qquad \epsilon > 0$$
$$(\mathbf{x}^L \pm \epsilon) \in S$$

Where ϵ is a small positive number. The point $\mathbf{x}^L$ is called a *local minimum*. A convex function has only a global minimum, while a nonconvex function may have an infinite number of local mimima. This property is of particular importance in most of the computational algorithms for the solution of minimization problems.

2.6 THE KUHN-TUCKER THEOREM [18]

One of the most fundamental theorems of mathematical programming, which serves as a basis of many computational algorithms, is the famous Kuhn-Tucker theorem [18]. A comprehensive exposition of the Kuhn-Tucker theorem can be found, for example, in references 1, chap. II; 2, chap. 6; and 3, chap. 3.

Let us restate the MP problem in the following form:

$$\begin{aligned} &\text{Minimize} \quad f(\mathbf{x}) \\ &\text{Subject to} \quad g_j(\mathbf{x}) \leqslant 0, \quad j = 1, \ldots, m \\ &\qquad\qquad\quad x_i \geqslant 0, \quad i = 1, \ldots, n \end{aligned} \qquad (2\text{-}22)$$

where $f(\mathbf{x})$ and $g_j(\mathbf{x})$ ($j = 1, \ldots, m$) are convex functions of n variables $\mathbf{x}^T = [x_1, x_2, \ldots, x_n]$.

Now introduce a set of m variables $u_1, \ldots, u_m$, usually called *Lagrange multipliers*, which form the vector $\mathbf{u}$. A new *Lagrange function* $L(\mathbf{x}, \mathbf{u})$ is defined as follows:

$$L(\mathbf{x}, \mathbf{u}) = f(\mathbf{x}) + \sum_{j=1}^{m} u_j g_j(\mathbf{x}) \qquad (2\text{-}23)$$

The Kuhn-Tucker theorem may now be stated:

Kuhn-Tucker theorem. A vector $\mathbf{x}^o$ is a solution to the optimization problem (2-22) if and only if a vector $\mathbf{u}^o$ exists such that

$$\begin{aligned} x_i^o &\geqslant 0, \quad i = 1, \ldots, n \\ u_j^o &\geqslant 0, \quad j = 1, \ldots, m \\ L(\mathbf{x}^o, \mathbf{u}) &\leqslant L(\mathbf{x}^o, \mathbf{u}^o) \leqslant L(\mathbf{x}, \mathbf{u}^o) \end{aligned} \qquad (2\text{-}24)$$

for all $x_i \geqslant 0$ and $u_j \geqslant 0$.

In other words, the optimal point $(\mathbf{x}^o, \mathbf{u}^o)$ has the following property: For a fixed $\mathbf{u}^o$, $L(\mathbf{x}, \mathbf{u}^o)$ has a global minimum over the domain $x_i \geqslant 0$ ($i = 1, \ldots, n$) at $(\mathbf{x}^o, \mathbf{u}^o)$. For a fixed $\mathbf{x}^o$, it has a global maximum over $u_j \geqslant 0$ ($j = 1, \ldots, m$) at the same point $(\mathbf{x}^o, \mathbf{u}^o)$. An extremal point with these properties is called a *saddle point*, and therefore the Kuhn-Tucker theorem is often referred to as the *saddle-point theorem*.

A heuristic argument to justify the theorem may be given as follows: Substituting Eq. (2-23) into (2-24), one obtains

$$f(\mathbf{x}^o) + \sum_{j=1}^{m} u_j g_j(\mathbf{x}^o) \leqslant f(\mathbf{x}^o) + \sum_{j=1}^{m} u_j^o g_j(\mathbf{x}^o) \leqslant f(\mathbf{x}) + \sum_{j=1}^{m} u_j^o g_j(\mathbf{x}) \qquad (2\text{-}25)$$

for all $x_i \geqslant 0$ ($i = 1, \ldots, n$) and $u_j \geqslant 0$ ($j = 1, \ldots, m$). From the left inequality of (2-25), it follows that

$$\sum_{j=1}^{m} u_j g_j(\mathbf{x}^o) \leqslant \sum_{j=1}^{m} u_j^o g_j(\mathbf{x}^o) \tag{2-26}$$

Since Eq. (2-26) is to hold for *all* $u_j \geqslant 0$, it follows that

$$g_j(\mathbf{x}^o) \leqslant 0 \tag{2-27}$$

and

$$\sum_{j=1}^{m} u_j^o g_j(\mathbf{x}^o) = 0 \tag{2-28}$$

As one may see, inequality (2-27) assures that the optimum point $\mathbf{x}^o$ is feasible. Taking into account Eq. (2-28), the right inequality of (2-25) becomes

$$f(\mathbf{x}^o) \leqslant f(\mathbf{x}) + \sum_{j=1}^{m} u_j^o g_j(\mathbf{x}) \tag{2-29}$$

for all $x_i \geqslant 0$.

Since $u_j^o \geqslant 0$ $(j = 1, \ldots, m)$ and $g_j(\mathbf{x}) \leqslant 0$, by feasibility requirements, it follows from Eq. (2-29) that

$$f(\mathbf{x}^o) \leqslant f(\mathbf{x}) \tag{2-30}$$

for all $x_i \geqslant 0$ $(i = 1, \ldots, n)$ and $g_j(\mathbf{x}) \leqslant 0$ $(j = 1, \ldots, m)$. That is, $f(\mathbf{x}^o)$ is indeed the minimum of $f(\mathbf{x})$ over the feasible domain, and $\mathbf{x}^o$ is the solution of (2-22). A rigorous proof of the Kuhn-Tucker theorem may be found in the original paper [18] and in the other references cited at the beginning of this section.

The Kuhn-Tucker conditions may also be expressed in the following form [2, chap. 6; 3, chap. 3]:

$$\left.\frac{\partial L}{\partial x_i}\right|_{\mathbf{x}^o, \mathbf{u}^o} \geqslant 0$$

$$x_i^o \left.\frac{\partial L}{\partial x_i}\right|_{\mathbf{x}^o, \mathbf{u}^o} = 0 \tag{2-31}$$

$$x_i^o \geqslant 0, \quad i = 1, \ldots, n$$

$$\left.\frac{\partial L}{\partial u_j}\right|_{\mathbf{x}^o, \mathbf{u}^o} \leqslant 0$$

$$u_j^o \left.\frac{\partial L}{\partial u_j}\right|_{\mathbf{x}^o, \mathbf{u}^o} = 0 \tag{2-32}$$

$$u_j^o \geqslant 0, \quad j = 1, \ldots, m$$

2.7 DUALITY

The concept of duality plays an important role in mathematical-programming methods.

Let us consider the same minimization problem as defined in Eq. (2-22):

$$(P) = \min \{f(\mathbf{x}) \mid g_j(\mathbf{x}) \leqslant 0, j = 1, \ldots, m; x_i \geqslant 0, i = 1, \ldots, n\} \quad (2\text{-}33)$$

This problem is denoted as the *primal problem* (P) [19, 20]. Corresponding to this problem, there is a related maximization problem known as the *dual* of the primal problem. The dual problem is formulated as follows:

$$(D) = \max \left\{ L(\mathbf{x}, \mathbf{u}) \left| \frac{\partial L(\mathbf{x}, \mathbf{u})}{\partial x_i} = 0, i = 1, \ldots, n; u_j \geqslant 0, j = 1, \ldots, m \right. \right\} \quad (2\text{-}34)$$

where $L(\mathbf{x}, \mathbf{u})$ is the Lagrange function defined in Eq. (2-23).

The concept of duality establishes certain relationships between the solutions of the primal problem and the dual problem. Theorem 2-1 describes the relation between the objective functions $f(\mathbf{x})$ and $L(\mathbf{x}, \mathbf{u})$ of the two problems.

Theorem 2-1. If $f(\mathbf{x})$ and $g_j(\mathbf{x})$ $(j = 1, \ldots, m)$ are convex functions, $\mathbf{y}$ is any feasible point of the primal problem of (2-33), and if $(\mathbf{x}^f, \mathbf{u}^f)$ denotes any feasible point of the dual problem of (2-34), then

$$f(\mathbf{y}) \geqslant L(\mathbf{x}^f, \mathbf{u}^f) \quad (2\text{-}35)$$

Proof. Since $\mathbf{y}$ is feasible,

$$g_j(\mathbf{y}) \leqslant 0, j = 1, \ldots, m$$

and therefore

$$f(\mathbf{y}) \geqslant f(\mathbf{y}) + \sum_{j=1}^{m} u_j^f g_j(\mathbf{y}) \quad (2\text{-}36)$$

Since both f and all g_j's are convex, the following expressions, which are alternative definitions of a convex function, may be written:

$$f(\mathbf{y}) \geqslant f(\mathbf{x}^f) + (\mathbf{y} - \mathbf{x}^f)^T \nabla f(\mathbf{x}^f) \quad (2\text{-}37)$$

$$g_j(\mathbf{y}) \geqslant g_j(\mathbf{x}^f) + (\mathbf{y} - \mathbf{x}^f)^T \nabla g_j(\mathbf{x}^f), \quad j = 1, \ldots, m \quad (2\text{-}38)$$

Substituting Eqs. (2-37) and (2-38) into Eq. (2-36), we get

$$f(\mathbf{y}) + \sum_{j=1}^{m} u_j^f g_j(\mathbf{y}) \geqslant f(\mathbf{x}^f) + \sum_{j=1}^{m} u_j^f g_j(\mathbf{x}^f)$$

$$+ (\mathbf{y} - \mathbf{x}^f)^T \left[\nabla f(\mathbf{x}^f) + \sum_{j=1}^{m} u_j^f \nabla g_j(\mathbf{x}^f) \right] \quad (2\text{-}39)$$

Since the point $(\mathbf{x}^f, \mathbf{u}^f)$ is dual feasible [feasible point of (D)],

$$\nabla_{\mathbf{x}} L(\mathbf{x}^f, \mathbf{u}^f) = \nabla f(\mathbf{x}^f) + \sum_{j=1}^{m} u_j^f \nabla g_j(\mathbf{x}^f) = 0 \quad (2\text{-}40)$$

One should also note that by definition,

$$f(\mathbf{x}^f) + \sum_{j=1}^{m} u_j^f g_j(\mathbf{x}^f) = L(\mathbf{x}^f, \mathbf{u}^f) \quad (2\text{-}41)$$

Considering Eq. (2-41) in light of Eqs. (2-36) and (2-39), we have

$$f(\mathbf{y}) \geqslant L(\mathbf{x}^f, \mathbf{u}^f) \quad \text{QED}$$

The practical impact of Theorem 2-1 consists of the fact that if dual feasible points are generated by algorithms that solve the primal problem (P), then a lowerbound is available on y^o which is the optimum value of $f(y)$. The function $f(y)$, which is being minimized, is the objective function of problem (P).

Before proceeding to Theorem 2-2, it is necessary to define the concept of Kuhn-Tucker's first-order constraint qualification [18].

Definition. Let x^f be a feasible point to the primal problem (P) and assume that the functions $g_j(x)$ $(j = 1, \ldots, m)$ are once differentiable. Then the *first-order constraint qualification* holds at x^f if for any non-zero vector z such that

$$z^T \nabla g_j(x^f) \leqslant 0 \text{ for all } j \in \{j \,|\, g_j(x^f) = 0\} \tag{2-42}$$

z is tangent to a once-differentiable arc, the arc emanating from x^f and contained in the constraint region.

Theorem 2-2. If the first-order constraint qualification holds at x^o, which is a solution to the primal problem (P), then there exists a solution to the dual problem (D), and the maximum value of L is equal to the minimum value of f for a primal feasible x. (A complete proof of this theorem is given in reference 19.)

Theorem 2-2 is very important in practical applications. It is possible that a certain MP problem is very difficult to handle numerically. On the other hand, its dual may be easier to solve. Theorem 2-2 assures that under certain conditions, the optimal values of the objective functions of the primal and the dual problems coincide. Thus it may be more convenient to solve the dual problem first and use the information obtained to solve the primal problem.

As a particular case, the concept of duality is formulated for a linear-programming problem [8]. Consider the following primal LP problem:

$$\max \{c^T x \,|\, Ax \leqslant b; \, x_i \geqslant 0, \, i = 1, \ldots, n\} \tag{2-43}$$

where all the variables are the same as in Eqs. (2-13) and (2-14).

The dual to this problem may be formulated as follows:

$$\min \{b^T u \,|\, A^T u \geqslant c^T; \, u_j \geqslant 0, \, j = 1, \ldots, m\} \tag{2-44}$$

The following theorems are of considerable importance [8].

Theorem 2-3. If x is any feasible solution to the primal problem in Eq. (2-43) and u is any feasible solution to the dual problem in Eq. (2-44), then

$$c^T x \leqslant b^T u \tag{2-45}$$

Proof. Since x is feasible to the primal,

$$(Ax)_i \leqslant b_i, \quad i = 1, \ldots, m \tag{2-46}$$

and since $u_i \geqslant 0$, $i = 1, \ldots, m$, for any dual feasible solution,

$$u_i(A\mathbf{x})_i \leqslant u_i b_i \tag{2-47}$$

Summing over i, Eq. (2-47) is written

$$\sum_{i=1}^{m} u_i(A\mathbf{x})_i \leqslant \sum_{i=1}^{m} u_i b_i \tag{2-48}$$

or

$$\mathbf{u}^T A \mathbf{x} \leqslant \mathbf{b}^T \mathbf{u} \tag{2-49}$$

Similarly, since $x_i \geqslant 0$ $(i = 1, \ldots, n)$, it follows from Eq. (2-44) that

$$\mathbf{u}^T A \mathbf{x} \geqslant \mathbf{c}^T \mathbf{x} \tag{2-50}$$

Comparing Eqs. (2-49) and (2-50), we have

$$\mathbf{c}^T \mathbf{x} \leqslant \mathbf{b}^T \mathbf{u} \quad \text{QED}$$

Theorem 2-4. If $\mathbf{x}^f$ is a primal feasible solution and $\mathbf{u}^f$ is a dual feasible solution such that

$$\mathbf{c}^T \mathbf{x}^f = \mathbf{b}^T \mathbf{u}^f \tag{2-51}$$

then $\mathbf{x}^f$ is an optimal solution to the primal problem in Eq. (2-43) and $\mathbf{u}^f$ is an *optimal* solution to the dual problem in Eq. (2-44).

Proof. From Eqs. (2-45) and (2-51) it follows that

$$\mathbf{c}^T \mathbf{x} \leqslant \mathbf{b}^T \mathbf{u}^f = \mathbf{c}^T \mathbf{x}^f \tag{2-52}$$

and since the primal is a maximization problem, $\mathbf{x}^f$ is the optimal solution. Similarly, for any feasible $\mathbf{u}$,

$$\mathbf{b}^T \mathbf{u}^f = \mathbf{c}^T \mathbf{x}^f \leqslant \mathbf{b}^T \mathbf{u} \tag{2-53}$$

and since the dual is a minimization problem, $\mathbf{u}^f$ is its optimal solution.

Theorem 2-5. If one of the problems in Eqs. (2-43) and (2-44) has an optimal solution, then the other also has an optimal solution.

A complete proof of Theorem 2-5 may be found in reference 8, pp. 229–230. It is easy to see that Theorems 2-3 and 2-4 are particular cases of the more general Theorems 2-1 and 2-2.

REFERENCES

1. J. Abadie, ed., *Nonlinear Programming*, Wiley, New York, 1967.
2. G. Hadley, *Nonlinear and Dynamic Programming*, Addison-Wesley, Reading, Mass., 1964.
3. H. P. Kunzi, W. Krelle, and W. Oettli, *Nonlinear Programming*, Blaisdell, Waltham, Mass., 1966.
4. S. Vajda, *Mathematical Programming*, Addison-Wesley, Reading, Mass., 1961.

5. R. L. Graves and P. Wolfe, eds., *Recent Advances in Mathematical Programming*, McGraw-Hill, New York, 1963.
6. G. Zoutendijk, *Methods of Feasible Directions*, American Elsevier, New York, 1960.
7. G. B. Dantzig, *Linear Programming and Extensions*, Princeton University Press, Princeton, N.J., 1963.
8. G. Hadley, *Linear Programming*, Addison-Wesley, Reading, Mass., 1962.
9. M. Simonnard, *Linear Programming*, Prentice-Hall, Englewood Cliffs, N.J., 1966.
10. S. I. Gass, *Linear Programming*, 3rd ed., McGraw-Hill, New York, 1969.
11. R. W. Llewellyn, *Linear Programming*, Holt, Rinehart and Winston, New York, 1964.
12. J. C. G. Boot, *Quadratic Programming*, North-Holland, Amsterdam, 1964.
13. R. J. Duffin, E. L. Peterson, and C. M. Zener, *Geometric Programming*, Wiley, New York, 1967.
14. D. J. Wilde and C. S. Beightler, *Foundation of Optimization*, Prentice-Hall, Englewood Cliffs, N.J., 1967.
15. M. L. Balinski, "Integer Programming: Methods, Uses, Computation," *Management Sci.*, **12**, pp. 253–313, 1965.
16. J. F. Benders, "Partitioning Procedures for Solving Mixed-Variables Programming Problems," *Numerische Math.*, **4**, pp. 238–252, 1962.
17. F. A. Valentine, *Convex Sets*, McGraw-Hill, New York, 1964.
18. H. W. Kuhn and A. W. Tucker, "Nonlinear Programming," *Proceedings of the Second Berkeley Symposium on Mathematical Statistics and Probability*, University of California Press, Berkeley, 1951, pp. 481–492.
19. A. V. Fiacco and G. P. McCormick, *Nonlinear Programming: Sequential Unconstrained Minimization Techniques*, Wiley, New York, 1968.
20. P. Wolfe, "A Duality Theorem for Nonlinear Programming," *Quart. Appl. Math.*, **19**, pp. 239–244, 1961.

3

NUMERICAL SOLUTION OF MATHEMATICAL-PROGRAMMING PROBLEMS

3.1 LINEAR PROGRAMMING

Linear-programming methods [1–5] have been applied to a variety of optimal-control problems, and in particular to discrete-data systems. In this section we shall reformulate the linear-programming problem by introducing the *slack variables*. This reformulation allows the convenient use of the simplex algorithm for the problem.

Reformulation of the Problem

As stated in Chapter 2, an LP problem can be formulated as follows: Given the objective function

$$f(\mathbf{x}) = \sum_{i=1}^{N} c_i x_i = \mathbf{c}^T \mathbf{x} \tag{3-1}$$

find the values of x_i, $i = 1, 2, \ldots, N$, satisfying the m linear inequality or equality constraints

$$\sum_{r=1}^{N} a_{ir}x_r \leqslant b_i, \qquad i = 1,\ldots,p \qquad (3\text{-}2)$$

$$\sum_{r=1}^{N} a_{jr}x_r = b_j, \qquad j = p+1,\ldots,p+q \qquad (3\text{-}3)$$

$$\sum_{r=1}^{N} a_{kr}x_r \geqslant b_k, \qquad k = p+q+1,\ldots,m \qquad (3\text{-}4)$$

That is, there are assumed to be q equality constraints and $m - q$ inequality constraints.

Since in general it is easier to work with equalities than inequalities, the inequalities of (3-2) and (3-4) are converted to equality form by introducing the *slack variables*.

For constraints of the form of (3-2), a slack variable $x_{si} \geqslant 0$ is introduced so that

$$\sum_{r=1}^{N} a_{ir}x_r + x_{si} = b_i, \qquad i = 1,\ldots,p \qquad (3\text{-}5)$$

For the constraints of (3-4), we can introduce the slack variable $x_{sj} \geqslant 0$ so that

$$\sum_{r=1}^{N} a_{jr}x_r - x_{sj} = b_j, \qquad j = p+q+1,\ldots,m \qquad (3\text{-}6)$$

It should be noted that slack variables do not appear in the objective function of (3-1) and, as will become clear later, that they do not influence the optimal solution. Therefore, in essence, the slack variables are used in order to bring all the constraints to the general form of

$$A\mathbf{x} = \mathbf{b} \qquad (3\text{-}7)$$

where $\mathbf{x}$ is now an $n = (N + m - q)$-dimensional vector which includes all the original and the slack variables, b is an m-dimensional constant vector, and A is an $m \times n$ constant matrix. Therefore, (3-7) represents m linear equations with n unknowns. It is assumed that $n > m$, i.e., that the total number of unknown variables is greater than the number of constraint equations. If $m > n$, there are more equations than unknowns, and some of the equations are dependent and, therefore, could be eliminated. If $m = n$, we have the usual case of solving a set of simultaneous linear algebraic equations, and a unique solution exists if A is nonsingular.

Now the LP problem may be stated in a compact form as follows:

$$\max \text{ (or min)} \{\mathbf{c}^T\mathbf{x} \,|\, A\mathbf{x} = \mathbf{b};\, x_i \geqslant 0,\, i = 1, 2, \ldots, n\} \qquad (3\text{-}8)$$

where

$$\mathbf{c}^T = [c_1, c_2, \ldots, c_N, 0, \ldots, 0]$$

is an n-dimensional row vector and N is the number of the original variables.

The Simplex Algorithm

One of the most efficient algorithms for the numerical solution of LP problems is the so-called *simplex algorithm*, originated by G. B. Dantzig [1]. There exist at present several versions of the simplex method. However, only the basic ideas of the algorithm will be outlined in this section. For a more detailed derivation and discussion of the simplex algorithm, the reader is referred to other texts on linear programming [1–5].

Rewriting the constraint equations in Eq. (3-7) in scalar form, we have

$$
\begin{aligned}
a_{11}x_1 + \cdots + a_{1n}x_n &= b_1 \\
&\vdots \\
a_{m1}x_1 + \cdots + a_{mn}x_n &= b_m
\end{aligned}
\tag{3-9}
$$

which is a set of m linear equations with n variables. It is assumed that $n > m$ and that the a_{ij}'s ($i = 1, \ldots, m; j = 1, \ldots, n$) are constant elements of the matrix A. As defined in Chapter 2, any vector $\mathbf{x}$ whose components, $x_j \geqslant 0$ ($j = 1, \ldots, n$), satisfy all the equations of Eq. (3-9) constitutes a *feasible solution* of the problem. In the simplex method, there is a particular form of a feasible solution which is of major importance, namely, a *basic feasible solution*.

Definition of a basic solution. Given a system of m simultaneous linear equations with n unknowns, as the one defined by Eq. (3-9), and assuming that the rank of matrix A is m, if any $m \times m$ nonsingular matrix is chosen from A, and if all the $n - m$ variables not associated with the columns of this matrix are set equal to zero, the solution to the resulting system of equations is called a *basic solution*. The m variables which can be different from zero are called the *basic variables*.

Naturally, a basic solution which is feasible is called a *basic feasible solution*.

Consider now the geometric structure of Eq. (3-9). Each one of the m equations constitutes a hyperplane in an n-dimensional space. It is easy to see that a hyperplane is a convex set. For example, suppose that points $\mathbf{x}_1$ and $\mathbf{x}_2$ are on the hyperplane; then

$$\mathbf{a}_i^T \mathbf{x} = b_i, \quad i = 1, \ldots, m \tag{3-10}$$

$$\mathbf{a}_i^T \mathbf{x}_1 = b_i, \tag{3-11}$$

and

$$\mathbf{a}_i^T \mathbf{x}_2 = b_i \tag{3-12}$$

Any point $\mathbf{x}$ located on the line segment joining the two points $\mathbf{x}_1$ and $\mathbf{x}_2$ is also on the same hyperplane, since

$$\mathbf{a}_i^T\mathbf{x} = \mathbf{a}_i^T[\lambda\mathbf{x}_2 + (1-\lambda)\mathbf{x}_1] = \lambda b_i + (1-\lambda)b_i = b_i \qquad (3\text{-}13)$$

The open and closed half-spaces bounded by the hyperplane are also convex sets. Suppose for a moment that the equations of Eq. (3-9) are inequalities. The feasible region would then be represented by the *intersection* of closed half-spaces, bounded by hyperplanes—or, in other words, an intersection of convex sets. Using the well-known theorem which states that the intersection of convex sets is also convex [6], one may state that the feasible region of an LP problem expressed by linear inequalities is a convex set. In the LP problem represented by equalities, Eq. (3-9), the feasible region constitutes the boundary of the convex set. At this point an important notion of the extreme points of convex sets should be introduced:

Definition of an extreme point. *A point* $\mathbf{x}$ *is an extreme point of a convex set if, and only if, there do not exist other points* $\mathbf{x}_1, \mathbf{x}_2, \mathbf{x}_1 \neq \mathbf{x}_2$ *in the set such that* $\mathbf{x} = \lambda\mathbf{x}_2 + (1-\lambda)\mathbf{x}_1$, $0 < \lambda < 1$.

That is, an extreme point cannot be located on a line segment joining any other two points of the set. An extreme point is actually a "corner point" of the convex set. For instance, in the example presented in Fig. 2-2, the extreme points of the convex set (polygon $ABCDO$) are A, B, C, D, and O.

By inspection of Fig. 2-2, one may see that a finite optimal solution is always an extreme point. Suppose that L_1 is the optimal solution. One may always move L_1 toward increasing values of L, along the edge BC, while remaining in the feasible region, until the extreme point B is reached. Moving L beyond the extreme point B would make the solution unfeasible. Hence, B is the point where the maximum value of the objective function is obtained without violating any of the constraints. Theorem 3-1 is of major importance:

Theorem 3-1. *Every basic feasible solution to the set of equations in Eq. (3-9) is an extreme point of the convex set of feasible solutions of Eq. (3-9), and conversely, every extreme point of Eq. (3-9) is a basic feasible solution* [2, pp. 101–102].

One could say that the simplex algorithm actually hinges on Theorem 3-1. The simplex algorithm includes the following steps:

1. Find any basic feasible solution (bfs) of the constraint equations of Eq. (3-9).
2. From the first bfs advance to the next bfs in the direction of the increasing value of the objective function.
3. Continue stepping from one bfs to another, until the highest value of the objective function is reached.

For a system of m equations with n variables ($n > m$), the total number of basic feasible solutions is equal to the total number of combinations of m items out of n:

$$C_n^m = \frac{n!}{m!(n-m)!} \tag{3-14}$$

In other words, this is the maximum number of the extreme points of the convex set, defined (or rather bounded) by the equations of Eq. (3-9). The extreme points, as pointed out before, are the candidates for the optimal solution of an LP problem. The iterations of the simplex algorithm consist of going from one extreme point to another in the direction of the increasing value of the objective function, and as one may see, the total number of possible iterations is bounded from above, and the optimal solution can be reached in a finite number of steps.

The Simplex Tableau

Now let us condiser the following LP problem:

$$\max \left\{ f(\mathbf{x}) = \sum_{i=1}^{N} c_i x_i \,\middle|\, \sum_{i=1}^{N} a_{ji} x_i \leqslant b_j; \right.$$
$$\left. b_j \geqslant 0, j = 1, \ldots, m; \, x_i \geqslant 0, i = 1, \ldots, N \right\} \tag{3-15}$$

By introducing m additional slack variables, $x_{N+1}, \ldots, x_{N+m}$, the inequality constraints are converted to

$$\sum_{i=1}^{N} a_{ji} x_i + x_{N+j} = b_j \tag{3-16}$$

$x_{N+j} \geqslant 0, j = 1, \ldots, m$. Or, in matrix form,

$$\mathbf{A}\mathbf{x} = \mathbf{b} \tag{3-17}$$

where

$$A = \begin{bmatrix} a_{11} & \cdots & a_{1N} & 1 & \cdots & 0 \\ \vdots & & \vdots & 0 & 1 & 0 \\ \vdots & & \vdots & & & \\ a_{m1} & \cdots & a_{mN} & 0 & 0 & 1 \end{bmatrix} \; m \times n \tag{3-18}$$

$$\longleftarrow N \longrightarrow | \longleftarrow m \longrightarrow$$

$n = N + m$, and

$$\mathbf{x} = \begin{bmatrix} x_1 \\ x_2 \\ \vdots \\ \vdots \\ x_N \\ \hdashline x_{N+1} \\ \vdots \\ \vdots \\ x_{N+m} \end{bmatrix} \begin{array}{l} \Big\} \text{original variables} \\ \\ \Big\} \text{slack variables} \end{array} \tag{3-19}$$

Inspecting Eq. (3-16) we see that the solution

$$x_1 = x_2 = \cdots = x_N = 0 \tag{3-20}$$

$$x_{N+j} = b_j, \quad j = 1, \ldots, m \tag{3-21}$$

is a *basic feasible solution*. It may, therefore, be used as a starting point in the solution of the LP problem. Using Eq. (3-16), we can express the basic variables x_{N+j} ($j = 1, \ldots, m$) in the following form:

$$x_{N+j} = b_j - \sum_{i=1}^{N} a_{ji} x_i \geqslant 0 \tag{3-22}$$

$j = 1, \ldots, m$.

The vector **c** of the objective function is augmented by m additional zero terms corresponding to the slack variables x_{N+j} ($j = 1, \ldots, m$). The new vector **c** is

$$\mathbf{c}^T = [c_1, \ldots, c_N, c_{N+1}, \ldots, c_{N+m}] \tag{3-23}$$

where

$$c_{N+1} = c_{N+2} = \cdots = c_{N+m} = 0 \tag{3-24}$$

The value of the objective function for the basic feasible solution of Eq. (3-20) is

$$f_0(\mathbf{x}) = \sum_{j=1}^{m} c_{N+j} x_{N+j} = \sum_{j=1}^{m} c_{N+j} b_j = 0 \tag{3-25}$$

In view of Eqs. (3-24) and (3-25), the objective function can be written as

$$f(\mathbf{x}) = \sum_{i=1}^{N} c_i x_i$$
$$= f_0(\mathbf{x}) + \sum_{i=1}^{N} \left(c_i - \sum_{j=1}^{m} a_{ji} c_{N+j} \right) x_i \tag{3-26}$$

or

$$f(\mathbf{x}) - f_0(\mathbf{x}) = \sum_{i=1}^{N} z_i x_i \tag{3-27}$$

where

$$z_i = c_i - \sum_{j=1}^{m} a_{ji} c_{N+j} = c_i \tag{3-28}$$

Notice that Eq. (3-27) denotes the difference between the objective function and that of the basic feasible solution.

Introducing the notation

$$b_j = a_{j0}, \quad j = 1, \ldots, m \tag{3-29}$$

an initial tableau for the simplex method is shown in Table 3-1.

It should be remembered that in the basic feasible solution the values of the first N variables, $x_1, \ldots, x_N$ (nonbasic), are zero, and that those of the basic variables, $x_{N+1}, \ldots, x_{N+m}$, are $b_1, \ldots, b_m$, respectively.

Suppose now that the following correction is contemplated:

Table 3-1

THE INITIAL TABLEAU

	b	x_1	x_2	$\cdots$	x_N	x_{N+1}	x_{N+2}	$\cdots$	x_{N+m}
x_{N+1}	a_{10}	a_{11}	a_{12}	$\cdots$	a_{1N}	1	0	$\cdots$	0
x_{N+2}	a_{20}	a_{21}	a_{22}	$\cdots$	a_{2N}	0	1	$\cdots$	0
$\vdots$	$\vdots$	$\vdots$	$\vdots$		$\vdots$	$\vdots$	$\vdots$		$\vdots$
x_{N+m}	a_{m0}	a_{m1}	a_{m2}	$\cdots$	a_{mN}	0	0	$\cdots$	1
$f(\mathbf{x})$	$f_0(\mathbf{x})$	z_1	z_2	$\cdots$	z_N	0	0	$\cdots$	0

Columns headed by $x_1, \ldots, x_N$ are Nonbasic variables; columns headed by $x_{N+1}, \ldots, x_{N+m}$ are Basic variables. Row labels $x_{N+1}, \ldots, x_{N+m}$ denote Basic variables.

The value of one of the nonbasic variables x_i ($i = 1, \ldots, N$) is changed from zero to one, while the other nonbasic variables remain zero. The nonzero basic variables are modified in such way that the solution remains feasible; that is, all the constraints are still satisfied. Then Eq. (3-27) gives

$$f(\mathbf{x}) - f_0(\mathbf{x}) = z_i, \quad i = 1, \ldots, N \qquad (3\text{-}30)$$

Therefore, z_i represents the amount by which the objective function $f(\mathbf{x})$ would change from that of the basic feasible solution if the correction described above is performed. If z_i is positive, it means that the value of $f(\mathbf{x})$ is increased as a result of the correction. Since we have a maximization problem, this correction will improve the solution and drive it in the direction of the optimum. If z_i is negative, the value of z will decrease as a result of the exchange. Following this discussion, we can perform the following test of the tableau:

Check all the z_i's ($i = 1, \ldots, N$). If one or more of the z_i's is positive, it means that the current solution is not yet optimal since it can be improved by the described exchange. If all the z_i's are negative, the optimal solution has been reached since it cannot be improved any further. If some of the z_i's are zero, a number of equivalent optimal solutions exist. For a more detailed treatment of this special case, the reader is referred to reference 11.

If more than one z_i is positive, we have a problem of deciding which one of the corresponding x_i's should be exchanged. One of the most important factors in this choice is the absolute value $|z_i|$. The greater it is, the greater will be the increase in the value of the objective function. Therefore, we tend to choose

$$\max |z_i|, \quad z_i > 0 \qquad (3\text{-}31)$$

There is, however, another condition to be considered in this case. Using

Eqs. (3-22) and (3-25), we have

$$f_0(\mathbf{x}) = \sum_{j=1}^{m} c_{N+j}\left(b_j - \sum_{i=1}^{N} a_{ji}x_i\right) \tag{3-32}$$

If for a specific x_i all the a_{ji}'s < 0 ($j = 1, \ldots, m$), $f_0(\mathbf{x})$ will grow indefinitely when x_i is increased arbitrarily, thus yeilding an infinite solution. So, in quest for a finite optimal solution we pose an additional criterion for the choice of x_i to be exchanged: At least one of the a_{ji}'s ($j = 1, \ldots, m$) is positive. If this condition is not satisfied for any of the x_i's for which $z_i > 0$, then the LP problem has an infinite solution.

Suppose that the nonbasic variable chosen to be exchanged is x_e. The next problem is to establish which one of the basic variables is to be taken out of the basis and placed instead of x_e among the nonbasic variables. The main consideration is this case is to make sure that the new solution is feasible, that is, that all the problem constraints are still satisfied after the exchange.

Since the vectors $\mathbf{x}_{N+j}$ ($j = 1, \ldots, m$) constitute a basis in the m-dimensional space considered, every other vector in this space can be expressed as a linear combination of these vectors. Taking into account that the form of $\mathbf{x}_{N+j}$ is

$$\mathbf{x}_{N+j} = \begin{bmatrix} 0 \\ \cdot \\ \cdot \\ \cdot \\ 0 \\ 1 \\ 0 \\ \cdot \\ \cdot \\ \cdot \\ 0 \end{bmatrix} \quad (j\text{th component})$$

the vector $\mathbf{a}_e$, which is the column of the tableau in Table 3-1 corresponding to x_e, can be expressed as

$$\mathbf{a}_e = \begin{bmatrix} a_{1e} \\ a_{2e} \\ \cdot \\ \cdot \\ \cdot \\ a_{me} \end{bmatrix} = \sum_{j=1}^{m} a_{je}\mathbf{x}_{N+j} \tag{3-33}$$

Suppose that the vector $\mathbf{x}_{N+r}$ is the one chosen to leave the basis. It is singled out in Eq. (3-33) in the following way:

$$\mathbf{a}_e = \sum_{\substack{j=1 \\ j \neq r}}^{m} a_{je} \mathbf{x}_{N+j} + a_{re} \mathbf{x}_{N+r} \qquad (3\text{-}34)$$

or

$$\mathbf{x}_{N+r} = \frac{1}{a_{re}} \mathbf{a}_e - \sum_{\substack{j=1 \\ j \neq r}}^{m} \frac{a_{je}}{a_{re}} \mathbf{x}_{N+j} \qquad (3\text{-}35)$$

Following Eqs. (3-9) and (3-10), we can also write

$$\mathbf{b} = \sum_{j=1}^{m} b_j \mathbf{x}_{N+j} = \sum_{\substack{j=1 \\ j \neq r}}^{m} b_j \mathbf{x}_{N+j} + b_r \mathbf{x}_{N+r} \qquad (3\text{-}36)$$

Substituting Eq. (3-35) into Eq. (3-36), we obtain

$$\mathbf{b} = \sum_{\substack{j=1 \\ j \neq r}}^{m} \left(b_j - b_r \frac{a_{je}}{a_{re}} \right) \mathbf{x}_{N+j} + \frac{b_r}{a_{re}} \mathbf{a}_e \qquad (3\text{-}37)$$

To maintain feasibility after the exchange, b_j ($j = 1, \ldots, m$) has to remain nonnegative. Using Eq. (3-37), this condition can be expressed as the following inequalities:

$$b_j - b_r \frac{a_{je}}{a_{re}} \geq 0 \qquad (3\text{-}38)$$

$$\frac{b_r}{a_{re}} \geq 0 \qquad (3\text{-}39)$$

Since $b_r \geq 0$, then $a_{re} > 0$. Therefore the row r, or $\mathbf{x}_{N+r}$, is to be chosen only among

$$a_{je} > 0, \qquad j = 1, \ldots, m \qquad (3\text{-}40)$$

This would assure that Eq. (3-39) is satisfied. Equation (3-38) can be rewritten in the following form:

$$\frac{b_j}{a_{je}} - \frac{b_r}{a_{re}} \geq 0 \qquad (3\text{-}41)$$

To assure the satisfaction of the inequality in Eq. (3-41) for all j, we choose r so that b_r/a_{re} is the smallest of all b_j/a_{je} ($j = 1, \ldots, m$). In other words, taking into account Eq. (3-40), the $\mathbf{x}_{N+j}$ vector to leave the basis is chosen according to the following criterion:

$$\min_j \left[\frac{b_j}{a_{je}}, \text{ for all } a_{je} > 0 \right] \qquad (3\text{-}42)$$

The element a_{re} at the intersection of the rth row, representing the $\mathbf{x}_{N+r}$ vector leaving the basis, and the eth column, representing the $\mathbf{x}_e$ vector entering the basis, is called the *pivot*. The rth row is called the *pivot row*, and the eth column is called the *pivot column*. Since vector $\mathbf{x}_e$ enters the basis, its column will now have the form

$$\begin{bmatrix} 0 \\ \vdots \\ 0 \\ 1 \\ 0 \\ \vdots \\ 0 \end{bmatrix} \quad (r\text{th pivot element}) \qquad (3\text{-}43)$$

To accomplish this, we have to do the following:

1. Divide all elements of the pivot row by a_{re}. This will yield $a'_{re} = 1$.
2. To obtain all other elements of the pivot column as zeros, we perform the following transformations:

$$a'_{ji} = a_{ji} - a_{ri}\frac{a_{je}}{a_{re}}, \qquad \begin{array}{l} j = 1, \ldots, m \\ j \neq r \end{array} \qquad (3\text{-}44)$$

$$b'_j = a'_{j0} = a_{j0} - a_{r0}\frac{a_{je}}{a_{re}}, \qquad \begin{array}{l} j = 1, \ldots, m \\ j \neq r \end{array} \qquad (3\text{-}45)$$

The new values of z'_i are computed as before [see Eq. (3-28)]:

$$z'_{[i]} = c_{[i]} - \sum_{[j]} a_{j[i]} c_j \qquad (3\text{-}46)$$

where $[i]$ = index extending over the nonbasic variables
$[j]$ = index extending over the basic variables

A new tableau is formed, as shown in Table 3-2. The new z'_i's are tested again, and the procedure continues in a similar manner until the optimal solution is reached. A simple example solved in detail as an illustration follows.

ILLUSTRATIVE EXAMPLE

Consider the example given in Sec. 2.4, Eqs. (2-15) through (2-17), which is graphically solved in Fig. 2-2. The same example will now be solved using the simplex tableau described earlier. The problem is

$$\max\{x_1 + 4x_2 \mid x_1 \leqslant 10, \, x_2 \leqslant 10, \, x_1 + x_2 \leqslant 14, \, x_1, x_2 \geqslant 0\} \qquad (3\text{-}47)$$

Introducing slack variables, the first three inequality constraints become

$$\begin{aligned} x_1 + x_2 + x_3 &= 14 \\ x_1 + x_4 &= 10 \\ x_2 + x_5 &= 10 \end{aligned} \qquad (3\text{-}48)$$

Table 3-2

THE SUBSEQUENT TABLEAU

		x_1	x_2	...	x_e	...	x_N	x_{N+1}	...	x_{N+r}	...	x_{N+m}
x_{N+1}	a'_{10}	a'_{11}	a'_{12}		0		a'_{1N}	1		a'_{1r}		0
x_{N+2}	a'_{20}	a'_{21}	a'_{22}		0		a'_{2N}	0		a'_{2r}		0
...	...	...	...		...		...	...		...		...
x_{N+r}	a'_{r0}	a'_{r1}	a'_{r2}		1		a'_{rN}	0		a'_{rr}		0
...	...	...	...		...		...	...		...		...
x_{N+m}	a'_{m0}	a'_{m1}	a'_{m2}		0		a'_{mN}	0		a'_{mr}		1
$z = f(\mathbf{x})$	$f'_0(\mathbf{x})$	z'_1	z'_2		0		z'_N	0		z'_r		0

In this example,

$$\mathbf{c}^T = [1\ 4] \quad \mathbf{b} = \begin{bmatrix} 14 \\ 10 \\ 10 \end{bmatrix}$$

$$A = \begin{bmatrix} 1 & 1 & 1 & 0 & 0 \\ 1 & 0 & 0 & 1 & 0 \\ 0 & 1 & 0 & 0 & 1 \end{bmatrix}$$

The initial basic variables are, natually, x_3, x_4, and x_5. The initial tableau is

Basic variables →			x_1	x_2	← Nonbasic variables
	x_3	14	1	1	
	x_4	10	1	0	
	x_5	10	0	$\boxed{1}$	← Pivot
	$f(\mathbf{x})$	$f_0(\mathbf{x}) = 0$	$z_1 = 1$	$z_2 = 4$	

Since the tableau columns corresponding to the basic variables constitute a unity matrix, there is no need to reproduce it every time. Since $c_3 = c_4 = c_5 = 0$,

$$f_0(\mathbf{x}) = \sum_{j=3}^{5} c_j x_j = 0$$

Using Eq. (3-28) we calculate

$$z_1 = c_1 = 1 > 0$$
$$z_2 = c_2 = 4 > 0$$

Since

$$|z_2| = 4 > |z_1| = 1$$

x_2 is chosen to enter the basis. Now,

$$\frac{b_1}{a_{12}} = 14 > \frac{b_3}{a_{32}} = 10$$

So, the third row is chosen as the pivot row; that is, x_5 will leave the basis and $a_{32} = 1$ is the pivot. The new tableau is calculated according to Eqs. (3-44) through (3-46):

Basic variables →			x_1	x_5	← Nonbasic variables
	x_3	4	$\boxed{1}$	0	
	x_4	10	1	0	
	x_2	10	0	1	
	$f(\mathbf{x})$	40	1	-4	

Now only $z_1 = 1 > 0$, so column 1 will be the pivot column and x_1 will enter the basis. We have

$$\frac{b'_1}{a'_{11}} = 4 < \frac{b'_2}{a'_{21}} = 10$$

so x_3 is chosen to leave the basis. The next tableau is

	x_3	x_5
x_1	4	1	0
x_4	6	0	0
x_2	10	0	1
$f(\mathbf{x})$	44	-1	-4

Since z_1, z_2 are nonpositive, the optimal solution of

$$x_1 = 4$$
$$x_2 = 10$$
$$z_{opt} = 44$$

is reached, which is the same solution as obtained in Fig. 2-2.

3.2 QUADRATIC PROGRAMMING

General Formulation

The general form of a quadratic-programming problem [7–11] was given in Chapter 2 in Eqs. (2-19) and (2-20). The problem will now be reformulated in more detail. Without loss of generality, a minimization problem will be considered. The objective function is

$$f(\mathbf{x}) = \mathbf{c}^T \mathbf{x} + \mathbf{x}^T D \mathbf{x} \tag{3-49}$$

where $\mathbf{x}$ = n-dimensional variables vector
$\mathbf{c}$ = n-dimensional constant vector
D = $n \times n$ constant matrix (additional properties of D will be established below)

Following the discussion in the beginning of Sec. 3.1, one may express the linear constraints of the problem as equalities,

$$A\mathbf{x} = \mathbf{b} \quad x_i \geqslant 0, \quad i = 1, \ldots, n \tag{3-50}$$

where $\mathbf{b}$ = m-dimensional constant vector
A = $m \times n$ constant matrix

Using the notation of Eq. (3-50), the vector **x** may also include slack variables if some of the original constraints are inequalities. Equation (3-50) may also be expressed as

$$\mathbf{g}(\mathbf{x}) = A\mathbf{x} - \mathbf{b} = \mathbf{0} \tag{3-51}$$

The QP problem can be written in a more condensed form:

$$\min_{\mathbf{x}} \{\mathbf{c}^T\mathbf{x} + \mathbf{x}^T D\mathbf{x} \,|\, \mathbf{g}(\mathbf{x}) = 0; \, x_i \geqslant 0; \, i = 1, \ldots, n\} \tag{3-52}$$

A natural procedure to solve this problem would be to apply the Kuhn-Tucker theorem, discussed in Sec. 2.6. One of the basic postulates of the Kuhn-Tucker theorem is the convexity of the objective function and of all the constraints. From Eqs. (3-51) and (3-52), one can see that $\mathbf{c}^T\mathbf{x}$ and $g(\mathbf{x})$ are linear and hence convex. It remains only to check the conditions that are to be satisfied by the matrix D for the quadratic form $\mathbf{x}^T D\mathbf{x}$ to be convex. This question is answered by Theorem 3-2:

Theorem 3.2 [7]. *A positive semidefinite quadratic form $\mathbf{x}^T D\mathbf{x}$ is a convex function.*

That is, given

$$\mathbf{x}^T D\mathbf{x} \geqslant 0 \quad \text{for all } \mathbf{x} \tag{3-53}$$

one would like to show that given any two points $\mathbf{x}_1$ and $\mathbf{x}_2$,

$$[\lambda\mathbf{x}_2 + (1-\lambda)\mathbf{x}_1]^T D[\lambda\mathbf{x}_2 + (1-\lambda)\mathbf{x}_1]$$
$$\leqslant \lambda\mathbf{x}_2^T D\mathbf{x}_2 + (1-\lambda)\mathbf{x}_1^T D\mathbf{x}_1 \tag{3-54}$$

for any $0 \leqslant \lambda \leqslant 1$.

Proof. The left-hand side of the inequality in Eq. (3-54) may be expressed as

$$[\mathbf{x}_1 + \lambda(\mathbf{x}_2 - \mathbf{x}_1)]^T D[\mathbf{x}_1 + \lambda(\mathbf{x}_2 - \mathbf{x}_1)]$$
$$= \mathbf{x}_1^T D\mathbf{x}_1 + 2\lambda(\mathbf{x}_2 - \mathbf{x}_1)^T D\mathbf{x}_1 + \lambda^2(\mathbf{x}_2 - \mathbf{x}_1)^T D(\mathbf{x}_2 - \mathbf{x}_1) \tag{3-55}$$

Since by the assumption of positive semidefiniteness either side of Eq. (3-55) is nonnegative, and since $0 \leqslant \lambda \leqslant 1$ (or $\lambda \geqslant \lambda^2$),

$$\lambda(\mathbf{x}_2 - \mathbf{x}_1)^T D(\mathbf{x}_2 - \mathbf{x}_1) \geqslant \lambda^2(\mathbf{x}_2 - \mathbf{x}_1)^T D(\mathbf{x}_2 - \mathbf{x}_1) \tag{3-56}$$

Using Eq. (3-56) in (3-55), one may obtain

$$[\mathbf{x}_1 + \lambda(\mathbf{x}_2 - \mathbf{x}_1)]^T D[\mathbf{x}_1 + \lambda(\mathbf{x}_2 - \mathbf{x}_1)]$$
$$\leqslant \mathbf{x}_1^T D\mathbf{x}_1 + 2\lambda(\mathbf{x}_2 - \mathbf{x}_1)^T D\mathbf{x}_1 + \lambda(\mathbf{x}_2 - \mathbf{x}_1)^T D(\mathbf{x}_2 - \mathbf{x}_1)$$
$$= \mathbf{x}_1^T D\mathbf{x}_1 + \lambda(\mathbf{x}_2 - \mathbf{x}_1)^T D\mathbf{x}_1 + \lambda(\mathbf{x}_2 - \mathbf{x}_1)^T D\mathbf{x}_2$$
$$= \lambda\mathbf{x}_2^T D\mathbf{x}_2 + (1 - \lambda)\mathbf{x}_1^T D\mathbf{x}_1 \tag{3-57}$$

Equation (3-57) is identical to Eq. (3-54). QED

As discussed in Sec. 2.5, convexity assures a global extremum. The Kuhn-Tucker theorem is one of the rigorous forms of expressing this fact.

Two computational algorithms for solving the QP problem will now be reviewed briefly. For a more detailed exposition and the presentation of additional techniques, the reader is referred to other sources [7–11].

Wolfe's Method [12]

In analogy with the expression in Eq. (2-23), let us write the Lagrange function $L(\mathbf{x}, \mathbf{u})$ for the QP problem expressed by Eq. (3-52):

$$L(\mathbf{x}, \mathbf{u}) = \mathbf{c}^T\mathbf{x} + \mathbf{x}^T D\mathbf{x} + \mathbf{u}^T \mathbf{g}(\mathbf{x}) \tag{3-58}$$

or in scalar form, following Eq. (2-20),

$$L(\mathbf{x}, \mathbf{u}) = \sum_{i=1}^{n} c_i x_i + \sum_{i=1}^{n}\sum_{j=1}^{n} x_i d_{ij} x_j + \sum_{j=1}^{m} u_j g_j(\mathbf{x}) \tag{3-59}$$

Since all the constraints $g_j(\mathbf{x})$ are of the equality type, one has a minimization problem amenable to the application of a method in analogy with the Lagrange multipliers, which are the u_j's in this case. And, indeed, for this case the Kuhn-Tucker conditions, given by Eq. (2-32), reduce to [9]

$$\left.\frac{\partial L}{\partial u_j}\right|_{\mathbf{x}^o, \mathbf{u}^o} = 0, \quad j = 1, \ldots, m \tag{3-60}$$

Equation (2-31) remains valid, of course, since the conditions of $x_i \geqslant 0$ have not been dropped.

Applying the first inequality of Eq. (2-31) to Eq. (3-58), one obtains, considering Eq. (3-51),

$$\frac{\partial L}{\partial \mathbf{x}} = \mathbf{c} + 2D\mathbf{x} + A^T \mathbf{u} \geqslant 0 \tag{3-61}$$

Applying Eq. (3-60) to Eq. (3-58), one again obtains the original equality constraints, as in Eq. (3-50). Now, introducing a new slack variable,

$$\mathbf{v} = \frac{\partial L}{\partial \mathbf{x}} \tag{3-62}$$

or

$$v_i = \frac{\partial L}{\partial x_i}, \quad i = 1, \ldots, n \tag{3-63}$$

Considering Eqs. (3-50), (3-61), (3-63), and (2-31), the QP problem in Eq. (3-52) may be rewritten as the following equivalent problem: Find the vector $\mathbf{x}^o$ which satisfies the following conditions:

$$A\mathbf{x} = \mathbf{b} \tag{3-64}$$

$$2D\mathbf{x} - \mathbf{v} + A^T \mathbf{u} = -\mathbf{c} \tag{3-65}$$

$$\mathbf{x}^T \mathbf{v} = 0 \tag{3-66}$$

$$x_i \geqslant 0$$

$$v_i \geqslant 0, \quad i = 1, \ldots, n$$

By the Kuhn-Tucker theorem and by the previous development, the vector $\mathbf{x}^o$, which is a solution of Eqs. (3-64) through (3-66), is also the optimal solution of the QP problem in Eq. (3-52). It should be noted that Eqs. (3-64) and (3-65) are *linear* in the variables $\mathbf{x}$, $\mathbf{v}$, and $\mathbf{u}$ and that they may be rewritten in the following matrix form:

$$\begin{bmatrix} A & 0 & 0 \\ 2D & -I_n & A^T \end{bmatrix} \begin{bmatrix} \mathbf{x} \\ \mathbf{v} \\ \mathbf{u} \end{bmatrix} = \begin{bmatrix} \mathbf{b} \\ -\mathbf{c} \end{bmatrix} \quad (3\text{-}67)$$

The dimensions of the submatrices in the left-hand matrix of Eq. (3-67) are

$A:$ $m \times n$

$2D:$ $n \times n$

$-I_n:$ $n \times n$ (a negative unit matrix)

The dimension of the whole matrix is $(m+n) \times (m+2n)$; i.e., one has a total of $m+n$ equations to solve for $m+2n$ variables. Since $x_i v_i = 0$ ($i = 1, \ldots, n$) and since all $x_i \geqslant 0$ ($v_i \geqslant 0$), only n out of the $2n$ variables $\mathbf{x}$, $\mathbf{v}$ may be different from zero. This means that only $n+m$ of the variables of the $n+m$ equations in Eq. (3-67) may be nonzero. Hence the only solutions of Eq. (3-67) that one should consider are the *basic solutions*. To identify the basic solutions of Eq. (3-67), the simplex method of linear programming may be directly applied. This constitutes the main idea of Wolfe's algorithm. For more details, the reader is referred to references 7, 9 (chap. 8), and 12.

The Frank and Wolfe Method

The Frank and Wolfe algorithm [13], which was developed prior to Wolfe's, starts from the same equations, i.e., Eqs. (3-64) and (3-65). The difference between this algorithm and Wolfe's algorithm is that in this case the third condition in Eq. (3-66) is not preserved initially. The algorithm tries to obtain a basic feasible solution to Eq. (3-67) without satisfying the condition

$$x_i v_i = 0, \quad i = 1, \ldots, n \quad (3\text{-}68)$$

which is the same as Eq. (3-66). If no feasible solution is obtained, the procedure is stopped. However, if Eq. (3-27) has a feasible solution, not necessarily satisfying Eq. (3-68), the original QP problem in Eq. (3-52) will have a bounded optimal solution [7]. After a basic feasible solution to Eq. (3-67) is obtained, the algorithm proceeds from one basic feasible solution to the other, using the simplex method, until the condition in Eq. (3-68) is satisfied, i.e., until the optimal solution is reached.

3.3 NONLINEAR PROGRAMMING

The Gradient Approach

As discussed in Sec. 2.4, a mathematical-programming problem with at least one nonlinear constraint is classified as a nonlinear-programming problem. On the other hand, even if all the constraints are linear except the objective function, one again has a NLP problem. Quadratic programming is, of course, a particular case of NLP. However, as may be seen from Sec. 3.2, applying the Kuhn-Tucker theorem to a QP problem reduces it to a LP problem, to which the simplex method may be applied. Because of this fact, QP problems are usually considered as a separate class.

The computational solution of a NLP problem is much more complicated. Unlike the LP or the QP cases (with convex objective function), there is no computational algorithm which would guarantee a solution to any NLP problem in a finite number of steps. A number of isolated algorithms have been developed in the past. Some of the algorithms utilize the so-called gradient approach, which is discussed in this section [7, chap. 9; 9, chaps. 11–14].

Initially, the case of linear constraints and a nonlinear objective function will be discussed.

The problem considered is

$$\max \{f(\mathbf{x}) \mid A\mathbf{x} \leqslant \mathbf{b}; \; x_i \geqslant 0, \; i = 1, \ldots, n\} \tag{3-69}$$

where $\mathbf{x} = n$-dimensional variables vector
$A = $ Constant $m \times n$ matrix
$\mathbf{b} = m$-dimensional constant vector
$f(\mathbf{x}) = $ Nonlinear, convex objective function of the x variables

As an illustration, a two-dimensional problem (n = 2) is presented in Fig. 3-1. The shaded region represents the feasible region, i.e., the set of all points $\mathbf{x} = (x_1, x_2)$ for which all the constraints are satisfied. The curves F_0 to F_5 represent the loci of constant values of the objective function $f(x_1, x_2)$ in an ascending order: $F_0 < F_1 < F_2 < F_3 < F_4 < F_5$. Points $O, E_1, \ldots, E_5$ are the extreme points of the feasible region. From Fig. 3-1, it is obvious that the extreme point E_3 represents the optimal solution, yielding the maximal value of the objective function,

$$f(\mathbf{x})_{\max} = F_4$$

while satisfying all the constraints. The question is, how should the optimal point, E_3, be reached through a computational algorithm, bearing in mind that $f(\mathbf{x})$ is nonlinear?

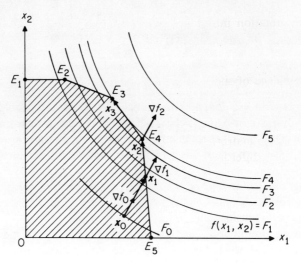

FIG. 3-1. Solution by a gradient approach.

One usually starts from a point $\mathbf{x}_0$ in the feasible region. Since it is a maximization problem, one would want to advance from $\mathbf{x}_0$ in the direction of increasing $f(\mathbf{x})$. Moreover, it is desirable to increase $f(\mathbf{x})$ as fast as possible. The gradient of $f(\mathbf{x})$,

$$\nabla f(\mathbf{x}) = \left| \frac{\partial f}{\partial x_1}, \frac{\partial f}{\partial x_2}, \ldots, \frac{\partial f}{\partial x_n} \right| \tag{3-70}$$

is a vector which points in the direction of the steepest increase in the value of this function. Therefore, the trajectory should be advanced from the initial point $\mathbf{x}_0$ in the direction of the gradient. On the other hand, since $f(\mathbf{x})$ is nonlinear, the direction of its gradient will change along the trajectory. Therefore, the direction of this advancement should be revised from time to time. One of the crucial problems in the gradient approach is to establish the optimal step size at each point considered. Many gradient techniques vary in the method of establishing this step size. As will be pointed out later, the direction of each step in the various gradient techniques is a function of ∇f but is not necessarily collinear with it. At this time, it may be appropriate to introduce a general notation which will be used in the following discussion of the gradient approach.

$\mathbf{x}_i =$ a point within the feasible region, reached after step i
$\nabla f_i =$ gradient of $f(\mathbf{x}_i) = \nabla f(\mathbf{x}_i) = \nabla f(\mathbf{x})|_{\mathbf{x}=\mathbf{x}_i}$
$\mathbf{r}_i =$ a unit vector in the direction of the step taken from the point $\mathbf{x}_i$; that is, $\mathbf{r}_i$ satisfies

$$\mathbf{r}_i^T \mathbf{r}_i = 1 \tag{3-71}$$

$d_i =$ a scalar, whose value is the step length taken from the point $\mathbf{x}_i$ in the direction of $\mathbf{r}_i$

Using the notation introduced above, the next point x_{i+1} from x_i is written

$$x_{i+1} = x_i + d_i r_i \qquad (3\text{-}72)$$

The dependence of the new direction vector r_i on the gradient ∇f_i is expressed as

$$r_i = H_i \nabla f_i \qquad (3\text{-}73)$$

where H_i is an $n \times n$ matrix, called the *direction matrix* or *metric* [14]. Various gradient techniques differ in the nature of the metric H_i. For instance, when

$$H_i = \frac{I_n}{|\nabla f_i|} \qquad (3\text{-}74)$$

where I_n is an n-dimensional unit matrix, the direction of the new step is collinear with the gradient ∇f_i at the starting point x_i. This technique is usually denoted as the *steepest ascent* method [14]. In this case, the step size d_i is chosen so that the objective function value at the next point, i.e., $f(x_{i+1})$, is maximized. Using Eqs. (3-72) through (3-74), we get

$$f(x_{i+1}) = f(x_i + d_i \nabla f_i) \qquad (3\text{-}75)$$

The function in Eq. (3-75) is maximized with respect to d_i. The necessary condition for the maximization is

$$\frac{df(x_{i+1})}{d(d_i)} = \nabla^T f_i \nabla f_{i+1} = 0 \qquad (3\text{-}76)$$

However, one must make sure that the next point x_{i+1} is still feasible. Equation (3-76) implies that the gradients at two successive points in the steepest ascent technique are *orthogonal*. This means that the path to the optimum solution is zigzagging at right angles from point to point, and usually at relatively small steps. The method of steepest ascent, which belongs to the so-called *small-step* gradient techniques, is in most cases very slow to converge [15]. More efficient techniques have been developed and will be discussed in the following sections.

The Method of Feasible Directions

The method of feasible directions, which belongs to the class of *large-step* gradient techniques, was developed by G. Zoutendijk [16]. The problem defined by Eq. (3-69) is considered. The $m \times n$ constraint matrix A is partitioned into row vectors:

$$A = \begin{bmatrix} a_1^T \\ \cdot \\ \cdot \\ \cdot \\ a_m^T \end{bmatrix} \qquad (3\text{-}77)$$

Then the constraints may be expressed in the following form:
$$\mathbf{a}_j^T \mathbf{x} \leq b_j, \quad j = 1, \ldots, m \tag{3-78}$$
Consider the problem of advancing from the point $\mathbf{x}_i$ to the next point $\mathbf{x}_{i+1}$, where $\mathbf{x}_i$ and $\mathbf{x}_{i+1}$ are related by
$$\mathbf{x}_{i+1} = \mathbf{x}_i + d_i \mathbf{r}_i \tag{3-79}$$
and $\mathbf{x}_{i+1}$ is feasible for sufficiently small d_i; i.e.,
$$\mathbf{a}_j^T \mathbf{x}_{i+1} \leq b_j \tag{3-80}$$
Since $\mathbf{x}_i$ is assumed feasible, it follows that the condition in Eq. (3-80) is equivalent to
$$\mathbf{a}_j^T \mathbf{r}_i \leq 0, \quad j \in J \tag{3-81}$$
where J is the set of all indices j for which
$$\mathbf{a}_j^T \mathbf{x}_i = b_j \tag{3-82}$$
A direction $\mathbf{r}_i$ satisfying the conditions in Eqs. (3-81) and (3-82) is called a *feasible direction*. The advance from $\mathbf{x}_i$ to $\mathbf{x}_{i+1}$ should be made in the direction of the increasing value of the objective function $f(\mathbf{x})$. Or, in other words, the inequality
$$\mathbf{r}_i^T \nabla f_i > 0 \tag{3-83}$$
should be satisfied. Zoutendijk [16] proposed several techniques of finding the feasible directions of iteration from the initial point $\mathbf{x}_i$. All the techniques consist of sequential solutions of intermediate mathematical-programming problems of the following nature.

In all the techniques, one uses the same objective function, which is maximized
$$\nabla f_i^T \mathbf{r}_i \tag{3-84}$$
and the constraints
$$\mathbf{a}_j^T \mathbf{r}_i \leq 0 \quad \text{for } j \in J \tag{3-85}$$
as defined by Eq. (3-81).

The various techniques denoted by T_i differ in the additional constraints imposed:
$$T_1: \quad \mathbf{r}_i^T \mathbf{r}_i \leq 1 \tag{3-86}$$
$$T_2: \quad -1 \leq r_{ik} \leq 1, \quad k = 1, \ldots, n \tag{3-87}$$
where r_{ik} is the kth component of $\mathbf{r}_i$
$$T_3: \quad \begin{array}{ll} r_{ik} \leq 1 & \text{for } \nabla f_{ik} > 0 \\ r_{ik} \geq -1 & \text{for } \nabla f_{ik} < 0 \end{array} \tag{3-88}$$
where ∇f_{ik} is the kth component of ∇f_i
$$T_4: \quad \nabla f_i^T \mathbf{r}_i \leq 1 \tag{3-89}$$
$$T_5: \quad \mathbf{a}_j^T (\mathbf{r}_i + \mathbf{x}_i) \leq b_j \quad \text{for all } j \tag{3-90}$$
The condition in Eq. (3-85) is actually included in Eq. (3-90).

Since the gradient at the point x_i, ∇f_i is specified, the objective function in Eq. (3-84) is linear. Inspecting all other constraints, one may see that techniques T_2 to T_5 yield LP problems. Only case T_1 in Eq. (3-86) represents a nonlinear constraint. Each linear subproblem may be solved by using the simplex algorithm. Once the feasible direction at the point x_i, r_i is found, the appropriate step size should be determined. This is accomplished by finding the following two candidates for this entry:

1. The value $d_i^{(1)}$ for which the path along the vector $d_i r_i$ leaves the feasible region defined by Eq. (3-78).
2. The value $d_i^{(2)}$ for which $f(x)$ attains its maximum along the vector $d_i r_i$—or the value $d_i^{(2)}$ for which the following equality is satisfied:

$$r_i^T \nabla f(x^i + d_i^{(2)} r_i) = 0 \tag{3-91}$$

[Compare with Eq. (3-76).]

After both values of $d_i^{(1)}$ and $d_i^{(2)}$ are calculated, the minimum value among them is determined as

$$d_i = \min(d_i^{(1)}, d_i^{(2)}) \tag{3-92}$$

One continues with the iteration procedure described above until a point x_p is reached at which

$$\max [\nabla f^T r_p] = 0 \tag{3-93}$$

that is, at which there is no new direction of advancement, r_p, which will maximize the value of the objective function $f(x)$.

Zoutendijk has also extended his method to problems with nonlinear constraints, such as

$$g_j(x) \leq b_j, \quad j = 1, \ldots, m \tag{3-94}$$

where the $g_j(x)$'s are assumed to be nonlinear convex functions. Suppose that the feasible point x_i has been reached at the last iteration. Let the class of all constraints for which

$$g_j(x_i) = b_j, \quad j \in J_1 \tag{3-95}$$

be denoted by J_1. The equation of a hyperplane which is tangent to each of the surfaces in Eq. (3-95) is given by

$$\nabla^T g_j(x_i) x = \nabla^T g_j(x_i) x_i \tag{3-96}$$

This type of hyperplane is referred to as the supporting hyperplane.

One has to find a new feasible point

$$x_{i+1} = x_i + d_i r_i \tag{3-97}$$

for which the objective function is increased; i.e.,

$$f(x_{i+1}) > f(x_i) \tag{3-98}$$

If x_{i+1} is feasible, i.e., if it satisfies

$$g_j(x_{i+1}) \leq b_j, \quad j \in J_1 \tag{3-99}$$

it is necessary that
$$\nabla g_j(\mathbf{x}_i)\mathbf{r}_i \leqslant 0 \tag{3-100}$$
Zoutendijk [16] suggested the following procedure for determining the next direction $\mathbf{r}_i$: An additional slack variable s is introduced and the following MP problem is solved. Maximize s subject to the constraints
$$\nabla g_j(\mathbf{x}_i)\mathbf{r}_i + s \leqslant 0, \quad j \in J_1 \tag{3-101}$$
$$-\nabla f(\mathbf{x}_i)\mathbf{r}_i + s \leqslant 0 \tag{3-102}$$
$$\mathbf{r}_i^T \mathbf{r}_i = 1 \tag{3-103}$$
As a result of solving this problem, a new direction $\mathbf{r}_i$ is found which keeps the trajectory in the feasible region while maintaining $\nabla f^T \mathbf{r}_i > 0$, i.e., which increases the value of the objective function. After finding $\mathbf{r}_i$, one establishes the maximum allowed step size d_m as the largest d_i for which
$$g_j(\mathbf{x}_i + d_i \mathbf{r}_i) \leqslant b_j \quad \text{for all } j \tag{3-104}$$
To establish the *actual* step size $0 < d_i < d_m$, one has to solve the problem
$$\max\{f(\mathbf{x}_i + d_i \mathbf{r}_i) \,|\, 0 \leqslant d_i \leqslant d_m\} \tag{3-105}$$
Additional details about Zoutendijk's method may be found in his book [16].

The Gradient Projection Method

The gradient projection method was developed by J. B. Rosen [17, 18]. Before going into the algorithm itself, the notion of the gradient projection is to be discussed. Consider the set of linear constraints as given by Eq. (3-78). On the boundaries of the feasible region, one would have the equality relations
$$\begin{aligned}\mathbf{a}_j^T \mathbf{x} &= b_j, \quad j = 1, \ldots, m \\ \mathbf{x} &= \{x_1, \ldots, x_n\}^T\end{aligned} \tag{3-106}$$
Each one of the m equations in Eq. (3-106) constitutes an $(n-1)$-dimensional *linear manifold* or *hyperplane* in the n-dimensional space. The intersection of q of Eq. (3-106) constitutes an $(n-q)$-dimensional linear manifold, denoted by Q. For $q = n$, Q would be a zero vector. For $q = 0$, Q would coincide with the entire n-dimensional space. Each one of the hyperplanes in Eq. (3-106) may be translated in parallel to the coordinate axes until it passes through the origin and obtains the form
$$\mathbf{a}_j^T \mathbf{x} = 0, \quad j = 1, \ldots, m \tag{3-107}$$
From Eq. (3-107) one can see that the vectors $\mathbf{a}_1, \ldots, \mathbf{a}_q$ are perpendicular to Q. On the other hand, since the vectors $\mathbf{a}_1, \ldots, \mathbf{a}_q$ are assumed to be linearly independent, they may serve as a basis of a q-dimensional linear manifold, denoted by $\mathbf{Q}$. If $q = n$, $\mathbf{Q}$ would coincide with the entire n-dimen-

sional space, and if $q = 0$, **Q** would consist of a zero vector only. Every vector **x** may be expressed as

$$\mathbf{x} = \sum_{j=1}^{q} u_j \mathbf{a}_j = A_q^T \mathbf{u} \tag{3-108}$$

where

$$A_q = \begin{bmatrix} \mathbf{a}_1^T \\ \cdot \\ \cdot \\ \cdot \\ \mathbf{a}_q^T \end{bmatrix} \qquad \mathbf{u} = \begin{bmatrix} u_1 \\ \cdot \\ \cdot \\ \cdot \\ u_q \end{bmatrix}$$

The two manifolds Q and **Q** are orthogonal and span the entire n-dimensional space, and every vector **x** may be expressed as a sum of two parts, $\mathbf{x}_Q$ and $\mathbf{x}_\varrho$; that is,

$$\mathbf{x} = \mathbf{x}_\mathbf{Q} + \mathbf{x}_\varrho \tag{3-109}$$

where

$$\mathbf{x}_\mathbf{Q} \in \mathbf{Q}$$
$$\mathbf{x}_\varrho \in Q$$

Since Q and **Q** are orthogonal,

$$\mathbf{x}_\varrho^T \mathbf{x}_\mathbf{Q} = 0 \tag{3-110}$$

Hence, if $\mathbf{x} \in \mathbf{Q}$, $\mathbf{x}_\mathbf{Q} = \mathbf{x}$ and $\mathbf{x}_\varrho = \mathbf{0}$; and if $\mathbf{x} \in Q$, $\mathbf{x}_\mathbf{Q} = \mathbf{0}$ and $\mathbf{x}_\varrho = \mathbf{x}$.

The vector $\mathbf{x}_\varrho$ is defined as the *projection* of **x** on Q and $\mathbf{x}_\mathbf{Q}$ as the projection of **x** on **Q**. The properties of the projection are

1. If $\mathbf{x}_\varrho \in Q$,

$$\mathbf{x}_\varrho^T (\mathbf{x} - \mathbf{x}_\varrho) = 0 \tag{3-111}$$

2.

$$|\mathbf{x}_\varrho - \mathbf{x}| < |\mathbf{y} - \mathbf{x}|, \quad \mathbf{y} \in Q, \quad \text{for all } \mathbf{y} \neq \mathbf{x}_\varrho \tag{3-112}$$

The projection $\mathbf{x}_\varrho$ may be expressed as a function of **x** using the projection matrix P_q:

$$\mathbf{x}_\varrho = P_q \mathbf{x} \tag{3-113}$$

$$P_q = I_n - A_q^T (A_q A_q^T)^{-1} A_q \tag{3-114}$$

We define

$$\mathbf{P}_q = I_n - P_q = A_q^T (A_q A_q^T)^{-1} A_q \tag{3-115}$$

Indeed, consider any vector **y** parallel to Q:

$$A_q \mathbf{y} = \mathbf{0} \tag{3-116}$$

Applying the projection matrix and using Eq. (3-116),

$$P_q \mathbf{y} = I_n \mathbf{y} - A_q^T (A_q A_q^T)^{-1} A_q \mathbf{y} = \mathbf{y} \tag{3-117}$$

$$\mathbf{P}_q \mathbf{y} = A_q^T (A_q A_q^T)^{-1} A_q \mathbf{y} = \mathbf{0} \tag{3-118}$$

Considering any vector $z \in Q$, so that by Eq. (3-108)

$$z = A_q^T u \tag{3-119}$$

and applying the projection matrix,

$$P_q z = A_q^T u - A_q^T (A_q A_q^T)^{-1} A_q A_q^T u$$
$$= A_q^T u - A_q^T u = 0 \tag{3-120}$$
$$\mathbf{P}_q z = A_q^T (A_q A_q^T)^{-1} A_q A_q^T u = A_q^T u = z \tag{3-121}$$

From the expressions in Eqs. (3-117), (3-118), (3-120), and (3-121), it follows that $P_q x$ is indeed the projection of x on Q; i.e.,

$$x_Q = P_q x \tag{3-122}$$

and that $\mathbf{P}_q x$ is the projection of x on $\mathbf{Q}$; i.e.,

$$x_Q = \mathbf{P}_q x \tag{3-123}$$

EXAMPLE

Consider the three-dimensional space shown in Fig. 3-2 ($n = 3$). The constraints corresponding to Eq. (3-106), with $q = 2$, are given as

$$\begin{aligned} x_1 &= 0 \\ x_2 &= 0 \end{aligned} \tag{3-124}$$

or

$$A_2 x = 0, \quad x = [x_1 \ x_2 \ x_3]^T$$

where

$$A_2 = \begin{bmatrix} 1 & 0 & 0 \\ 0 & 1 & 0 \end{bmatrix}$$

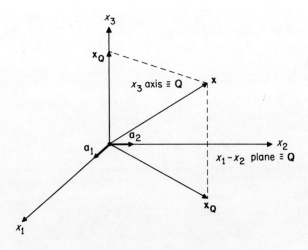

FIG. 3-2. Projections in three-dimensional space.

The intersection of the two manifolds given by Eq. (3-124) is the x_3 axis. The vectors $\mathbf{a}_1$ and $\mathbf{a}_2$, where

$$\mathbf{a}_1^T = [1 \quad 0 \quad 0]$$

and

$$\mathbf{a}_2^T = [0 \quad 1 \quad 0]$$

span the two-dimensional space $\mathbf{Q}$, which is the x_1-x_2 plane. In this case, the projection matrices are

$$\mathbf{P}_2 = A_2^T(A_2 A_2^T)^{-1} A_2 = \begin{bmatrix} 1 & 0 & 0 \\ 0 & 1 & 0 \\ 0 & 0 & 0 \end{bmatrix}$$

$$\overline{\mathbf{P}}_2 = I_3 - \mathbf{P}_2 = \begin{bmatrix} 0 & 0 & 0 \\ 0 & 0 & 0 \\ 0 & 0 & 1 \end{bmatrix}$$

Now one may return to the solution of the NLP problem. Suppose that at the end of the ith iteration, one has reached the point $\mathbf{x}_i$ at which the following set of constraints is satisfied:

$$\mathbf{a}_j \mathbf{x}_i = b_j, \qquad j = 1, \ldots, q \tag{3-125}$$

In other words, point $\mathbf{x}_i$ belongs to the intersection Q of the manifolds given by Eq. (3-125). As in any solution, the next direction $\mathbf{r}_i$ must be determined. The main idea of Rosen's method is that $\mathbf{r}_i$ is taken in the direction of the projection of the gradient ∇f_i on Q. Therefore,

$$\mathbf{r}_i = \alpha P_q \nabla f_i \tag{3-126}$$

where α is a proportionality constant and P_q is given by Eq. (3-114). The expression $P_q \nabla f_i$ is the so-called *gradient projection*. The step size d_i is chosen in the usual manner under the conditions that $f(\mathbf{x})$ is to continue to increase along the path and that the path remains within the feasible region. If the gradient projection $P_q \nabla f_i$ is zero, it means that the gradient ∇f_i belongs to the manifold $\mathbf{Q}$, spanned by the vectors $\mathbf{a}_1, \ldots, \mathbf{a}_q$, as shown by Eq. (3-120). Consequently, ∇f_i may be expressed as a linear combination of those base vectors

$$\nabla f_i = \sum_{j=1}^{q} u_j \mathbf{a}_j \tag{3-127}$$

where the u_j's $(j = 1, \ldots, q)$ are scalar coefficients. It should be remembered that $\mathbf{Q}$ and Q are orthogonal. In view of this, and if $u_j \geqslant 0$ $(j = 1, \ldots, q)$, the Kuhn-Tucker conditions in Eqs. (2-31) and (2-32) would be satisfied, and $\mathbf{x}_i$ would then be the optimal point. If some of the u_j's are negative, one deletes from the Q manifold the constraints associated with them, i.e., the corresponding $\mathbf{a}_j$, and proceeds in the same manner.

In the case of nonlinear constraints, one takes the gradient projection on the manifold formed by the tangent hyperplanes at the point $\mathbf{x}_i$. To ensure feasibility, a correction procedure in the direction of the feasible region is undertaken. This usually consists of a considerable number of iterations. For more details, the reader is referred to reference 18.

The Sequential Unconstrained Minimization Technique

The sequential unconstrained minimization technique (SUMT) was originally proposed by C. W. Carroll [19] and then rigorously justified and developed into a working computer program by A. V. Fiacco and G. P. McCormick [20]. One of the main advantages of this method is that it deals efficiently with nonlinear objective functions and constraints of both equality and inequality types. The convexity or concavity of the functions involved is not a necessary condition for the algorithm to work. However, if those conditions are not satisfied, the algorithm may converge to a local instead of a global extremum.

The problem considered is identical to the general MP problem, as expressed in Eq. (2-6). It is repeated here for the sake of convenience:

$$\min \{f(\mathbf{x}) \,|\, g_i(\mathbf{x}) \geqslant 0, \, i = 1, \ldots, q;\, h_j(\mathbf{x}) = 0, \, j = 1, \ldots, p\} \quad (3\text{-}128)$$

The basic idea of the SUMT method is the transformation of the constrained MP problem in Eq. (3-128) into a series of unconstrained minimization problems by using certain penalty functions for the constraints. To achieve this, one defines a modified objective function of the following form:

$$P(\mathbf{x}, r) = f(\mathbf{x}) - r \sum_{i=1}^{q} \ln g_i(\mathbf{x}) + r^{-1} \sum_{j=1}^{p} h_j^2(\mathbf{x}) \quad (3\text{-}129)$$

where r is a weighting coefficient and $r^{-1} = 1/r$. The SUMT algorithm proceeds in the following way:

1. An initial point $\mathbf{x}_0$ within the region defined by the inequality constraints is chosen and

$$g_i(\mathbf{x}_0) > 0, \quad i = 1, \ldots, q \quad (3\text{-}130)$$

2. An initial value of r, say $r_0 = 1$, is chosen. The actual value of r_0 is immaterial; one may start practically with any r_0, provided that it does not make the last two terms of Eq. (3-129) too small at the beginning.
3. The function $P(\mathbf{x}, r_0)$ is minimized, starting from point $\mathbf{x}_0$ in the direction of decreasing the value of $P(\mathbf{x}, r_0)$. The minimum of $P(\mathbf{x}, r_0)$ is assumed to occur at point $\mathbf{x}_1$. A variety of methods of unconstrained minimization may be used. Some of these methods will be discussed in Sec. 3.4.
4. A new value of r is computed in the following way:

$$r_1 = \frac{r_0}{K} \quad K > 1 \quad (3\text{-}131)$$

Usually one chooses K in the neighborhood of 4 or 5; however, this choice is not critical. Experience has shown that it is not advantageous to choose a K which is too large, say $K > 10$, since it tends to decrease the last terms of $P(\mathbf{x}, r)$ too much prematurely. The new coefficient r_1 is substituted into Eq. (3-129), and $P(\mathbf{x}, r_1)$ is minimized starting from point $\mathbf{x}_1$ in the direction of the decreasing value of $P(\mathbf{x}, r_1)$.

5. Step 4 is repeated for smaller and smaller values of r, obtained by the formula

$$r_{k+1} = \frac{r_k}{K} \qquad (3\text{-}132)$$

The computation is terminated when the final convergence criterion is satisfied; i.e., when

$$|P(\mathbf{x}, r_{k-1}) - P(\mathbf{x}, r_k)| < \epsilon, \quad \epsilon > 0 \qquad (3\text{-}133)$$

where ϵ is a prescribed small number depending on the precision required.

Inspecting Eq. (3-129), one can see that as r becomes very small and the constraints of the problem remain satisfied,

$$P_{\min}(\mathbf{x}, r) = f_{\min}(\mathbf{x}) + \delta \qquad (3\text{-}134)$$

where δ is a very small number which may be made as small as desired by specifying more stringent convergence criteria. To see that Eq. (3-134) is indeed correct, one may argue as follows.

As was pointed out, the algorithm always proceeds along points $\mathbf{x}_k$ inside the region defined by $g_i(\mathbf{x}) > 0$, $i = 1, \ldots, q$. If even one of the inequality constraints $g_i(\mathbf{x})$ comes too close to the feasibility boundary, i.e., $g_i(\mathbf{x}) \to 0$, one can see from Eq. (3-129) that $P(\mathbf{x}, r) \to \infty$, which is contrary to the minimization procedure. Therefore, as long as $P(\mathbf{x}, r)$ is being minimized, the inequality constraints will be satisfied automatically. Eventually, as r approaches zero, the second term on the right hand side of Eq. (3-129) will have a negligible value. On the other hand, r^{-1} becomes a very large number as r decreases. By minimizing $P(\mathbf{x}, r)$, the positive functions $h_j^2(\mathbf{x})$ are naturally driven to zero to satisfy eventually the equality constraints $h_j(\mathbf{x}) = 0, j = 1, \ldots, p$, as well. Therefore, one can see that by sequentially minimizing the functions $P(\mathbf{x}, r)$ for decreasing values of r, one should naturally obtain convergence to the minimum of $f(\mathbf{x})$ subject to all the specified constraints.

A rigorous proof of the convergence of the method when $f(\mathbf{x})$ and all $h_j^2(\mathbf{x})$'s are convex and all $g_i(\mathbf{x})$'s are concave has been given by Fiacco and McCormick [20]. An efficient program, also available on SHARE 3189, has been developed by Fiacco and McCormick.

3.4 UNCONSTRAINED MINIMIZATION METHODS

Introduction

Some of the techniques described in Sec. 3.3, and in particular the SUMT method, require numerous intermediate solutions of unconstrained extremization problems. One of the possibilities of performing unconstrained extremization is through direct search. Many efficient search methods have been developed [21; 22, chap. 6]. However, there are considerable difficulties in applying these methods in multivariable problems. It is much more efficient to apply methods based on the gradient approach, discussed in Sec. 3.3. Therefore, in this section attention will primarily be given to those methods. The various search methods will not be discussed since there exist two excellent texts covering this subject to which the reader is once again referred [21; 22, chap. 6].

Variable Metric Methods

The variable metric methods, which belong to the gradient approach type, were originally contributed by numerous authors. Recently, a comprehensive summary of these methods was reported by J. D. Pearson [14], and several computational evaluations were performed by other authors [14, 23–25].

The same notation introduced in Sec. 3.3 in conjunction with the gradient approach will also be used here. In addition, the following auxiliary entities will be defined.

s_i = the actual step vector from point x_i:

$$s_i = x_{i+1} - x_i = d_i r_i = d_i H_i \nabla f_i \tag{3-135}$$

y_i = the change in the value of the gradient from point x_i to point x_{i+1}:

$$y_i = \nabla f_{i+1} - \nabla f_i \tag{3-136}$$

$S_i = [s_0 \; s_1 \; \cdots \; s_{i-1}]$, an $n \times i$ matrix whose columns are the step vectors $s_0, s_1, \ldots, s_{i-1}$.

$Y_i = [y_0 \; y_1 \; \cdots \; y_{i-1}]$, an $n \times i$ matrix whose columns are the vectors $y_0, y_1, \ldots, y_{i-1}$.

As mentioned in Sec. 3.3, the matrix H_i, which is called the *metric*, establishes the dependence of the direction vector r_i on the gradient ∇f_i, as may be seen from Eqs. (3-73) and (3-135). For instance, in the methods of steepest ascent, $H_i = I/|\nabla f_i|$ for all i. In the variable metric methods, the metric H_i is changed from step to step. The various techniques depend on the actual algorithm according to which the H_i is changed. Once the new H_i's, and consequently the r_i's, have been established, one finds the step size d_i by a one-dimensional minimization (or maximization) process:

$$f(\mathbf{x}_{i+1}) = \min_d [f(\mathbf{x}_i + d\mathbf{r}_i)] \qquad (3\text{-}137)$$

The problem expressed by Eq. (3-137) may be solved by one of the search techniques in reference 21 or 22. The recursive formulas for H_i, used in various techniques, are listed.

1. Generalized Newton-Raphson Method [23]

$$H_i = [A(\mathbf{x}_i)]^{-1} \qquad (3\text{-}138)$$

where $A(x)$ is the *Hessian matrix* defined by

$$A(\mathbf{x}) = \begin{bmatrix} \frac{\partial^2 f}{\partial x_1^2} & \frac{\partial^2 f}{\partial x_1 \partial x_2} & \cdots & \frac{\partial^2 f}{\partial x_1 \partial x_n} \\ \cdot & \frac{\partial^2 f}{\partial x_2^2} & \cdots & \\ \cdot & \cdot & \ddots & \\ \frac{\partial^2 f}{\partial x_n \partial x_1} & \cdot & & \frac{\partial^2 f}{\partial x_n^2} \end{bmatrix} \qquad (3\text{-}139)$$

This method is most effective and fast-converging for a large class of problems. There may be difficulties, however, if A turns out to be nonpositive definite in a minimization problem or if A^{-1} does not exist. Fiacco and McCormick have proposed a modification of this method which overcomes some of the mentioned difficulties [20, pp. 166–167]. According to the modified Newton-Raphson procedure, if A has a negative eigenvalue, $\mathbf{r}_i$ is chosen to satisfy

$$\mathbf{r}_i^T A \mathbf{r}_i < 0 \qquad (3\text{-}140)$$

and

$$\mathbf{r}_i^T \nabla f_i \leqslant 0 \qquad (3\text{-}141)$$

If all the eigenvalues of A are greater than or equal to zero, $\mathbf{r}_i$ is chosen to satisfy

$$A\mathbf{r}_i = 0$$
$$\mathbf{r}_i^T \nabla f_i < 0 \qquad (3\text{-}142)$$

or

$$A\mathbf{r}_i = -\nabla f_i \qquad (3\text{-}143)$$

The Newton-Raphson technique belongs to the class of so-called second-order gradient methods. Although it has its own difficulties and requires more computer storage than the first-order methods (i.e., the ones utilizing first-order derivatives only), it is still the most efficient for a certain class of problems [25].

2. Projection Algorithm [14, 16, 23]

$$H_0 = R$$
$$H_{i+1} = H_i - \frac{(H_i y_i)(H_i y_i)^T}{y_i^T H_i y_i} \tag{3-144}$$

where R is a symmetrical positive definite matrix. After every n steps, H_i is reset back to R.

3. Fletcher-Powell-Davidon Method

This method, originally proposed by W. C. Davidon [26] and subsequently modified by R. Fletcher and M. J. D. Powell [27], works according to the following recursive formula:

$$H_0 = R$$
$$H_{i+1} = H_i - \frac{(H_i y_i)(H_i y_i)^T}{y_i^T H_i y_i} + \frac{s_i s_i^T}{s_i^T y_i} \tag{3-145}$$

where R is a symmetrical positive definite matrix. McCormick and Pearson [23] introduced the option of resetting H_i to R after every $n + 1$ steps. For the same problems it speeded up the solution by a factor of 4 to 5 [23]. It is interesting to note that as $i \to \infty$, $H_i \to [A(x_i)]^{-1}$, the Hessian matrix used in the Newton-Raphson method. Therefore, it could also be classified as a *quasi-Newton-Raphson* method, since its metric serves as an approximation of the Hessian matrix.

Both the projection and Fletcher-Powell-Davidon algorithms belong to the general class of *conjugate-direction* methods [28]. The directions r_0, $r_1, \ldots, r_{n-1}$ are said to be conjugate with respect to a positive definite symmetrical matrix A if

$$\begin{aligned} r_i^T A r_j &= 0, \quad 0 \leqslant i \neq j \leqslant n - 1 \\ r_i^T A r_i &> 0, \quad 0 \leqslant i \leqslant n - 1 \end{aligned} \tag{3-146}$$

Additional algorithms and reports on computational experimentation with conjugate-gradient methods may be found in other references [23, 24, 29–31]. The description of other variable metric algorithms and some experimental results and applications may be found in references 14, 20, 23–25, and 29.

A comprehensive presentation of various unconstrained minimization techniques may be found in the book by Fiacco and McCormick [20, chap. 7 and 8].

A Comparative Study [25]

To illustrate the relative efficiency of various unconstrained minimization techniques, a comparative study is made in the following. The unconstrained minimization methods tested are used in conjunction with the

Table 3-3 [14, 25]†

Method	Recursive Formula	Number of Iterations	Computing Time (sec)
1. Newton-Raphson	$H_i = A^{-1}(\mathbf{x}_i)$	61	1.40
2. Reduced gradient projection	$\mathbf{r}_{i+1} = \begin{bmatrix} -Y_1^{-1}Y_2 \\ I_n \end{bmatrix} \begin{bmatrix} -Y_1^{-1}Y_2 \\ I_n \end{bmatrix}^T \nabla f_{i+1}; \; Y_{i+1}r_{i+1} = 0$	114	2.10
3. Projected gradient	$H_{i+1} = H_i + \dfrac{(H_i y_i)(H_i y_i)^T}{\mathbf{y}_i^T H_i \mathbf{y}_i}$	156	2.93
4. Unsymmetrical variable metric 1	$H_{i+1} = H_i + \dfrac{(\mathbf{s}_i - H_i \mathbf{y}_i)(H_i^T \mathbf{y}_i)^T}{\mathbf{y}_i^T H_i \mathbf{y}_i}$	344	6.13
5. Modified Fletcher-Powell	$H_{i+1} = H_i + \dfrac{(\mathbf{s}_i - H_i \mathbf{y}_i)(\mathbf{s}_i - H_i \mathbf{y}_i)^T}{\mathbf{y}_i^T(\mathbf{s}_i - H_i \mathbf{y}_i)}$	354	6.63
6. Fletcher-Reeves	$\mathbf{r}_{i+1} = -\nabla f_{i+1} + \mathbf{r}_i \dfrac{\nabla f_{i+1}^T \nabla f_{i+1}}{\nabla f_i^T \nabla f_i}$	399	7.66
7. Unsymmetrical variable metric 2	$H_{i+1} = H_i + \dfrac{(\mathbf{s}_i - H_i \mathbf{y}_i)\mathbf{s}_i^T}{\mathbf{s}_i^T \mathbf{y}_i}$	580	9.67
8. Fletcher-Powell-Davidon	$H_{i+1} = H_i - \dfrac{(H_i \mathbf{y}_i)(H_i \mathbf{y}_i)^T}{\mathbf{y}_i^T H_i \mathbf{y}_i} + \dfrac{\mathbf{s}_i \mathbf{s}_i^T}{\mathbf{s}_i^T \mathbf{y}_i}$	2015	32.24
9. Steepest descent	$H_i = I_n/|\nabla f_i|$	4141	57.53

†In the majority of the methods, $H_0 = I_n$, and H_i is being reset to H_0 after every $n - 1$, or n, or $n + 1$ steps.

SUMT code (see Sec. 3.3). The particular problem tested is a NLP problem formulated for the fourth-order system discussed in Sec. 6.4, Eqs. (6-41) through (6-44).

Denoting the problem variables as

$$
\begin{array}{cc}
x & x_1 \\
q & x_2 \\
\delta_1 & x_3 \\
\omega_{n1} & x_4 \\
\delta_2 & x_5 \\
\omega_{n2} & x_6
\end{array}
$$

the NLP problem becomes the following: Minimize

$$J = cx_1 + x_3 + x_5$$

subject to the constraints

$$(2x_3^2 - 1)x_4^2 + (2x_5^2 - 1)x_6^2 \geqslant 0$$

$$x_1[4x_4^2 x_6^2(4x_3^2 x_5^2 - 2x_3^2 - 2x_5^2 + 1) + x_4^4 + x_6^4]$$
$$+ 1 - 2x_2(x_3 x_4 + x_5 x_6) \geqslant 0$$

$$2x_1 x_4^2 x_6^2[(2x_3^2 - 1)x_6^2 + (2x_5^2 - 1)x_4^2]$$
$$+ 2x_4 x_6 x_2(x_3 x_6 + x_5 x_4) - x_4^2 - x_6^2 - 4x_3 x_5 x_4 x_6 \geqslant 0$$

$$x_{3\,\min} \leqslant x_3 \leqslant x_{3\,\max}$$
$$x_{5\,\min} \leqslant x_5 \leqslant x_{5\,\max}$$
$$0 \leqslant x_4 \leqslant x_{4\,\max}$$
$$0 \leqslant x_6 \leqslant x_{6\,\max}$$

Notice that although the objective function is linear, some of the constraints are highly nonlinear. The computations are performed on the IBM 360/91 System, and the results are tabulated in Table 3-3.

The reader should be cautioned not to interpret these results in a universal manner. They mean only that for the particular type of problem under consideration, the Newton-Raphson technique proved to be the most effective. For a different problem, the results may be different.

REFERENCES

1. G. B. Dantzig, *Linear Programming and Extensions*, Princeton University Press, Princeton, N.J., 1963.
2. G. Hadley, *Linear Programming*, Addison-Wesley, Reading, Mass., 1962.
3. M. Simonnard, *Linear Programming*, Prentice-Hall, Englewood Cliffs, N.J., 1966.

4. S. I. Gass, *Linear Programming*, 3rd ed., McGraw-Hill, New York, 1969.
5. R. W. Llewellyn, *Linear Programming*, Holt, Rinehart and Winston, New York, 1964.
6. F. A. Valentine, *Convex Sets*, McGraw-Hill, New York, 1964.
7. G. Hadley, *Nonlinear and Dynamic Programming*, Addison-Wesley, Reading, Mass., 1964, chap. 7.
8. J. Abadie, ed., *Nonlinear Programming*, Wiley, New York, 1967, chap. VII by E. M. L. Beale.
9. M. P. Kunzi, W. Krelle, and W. Oettli, *Nonlinear Programming*, Blaisdell, Waltham, Mass., 1966, chaps. 4–10.
10. J. C. G. Boot, *Quadratic Programming*, North-Holland, Amsterdam, 1964.
11. H. P. Kunzi, H. G. Tzschach, and C. A. Zehnder, *Numerical Methods of Mathematical Optimization*, Academic Press, New York, 1968.
12. P. Wolfe, "The Simplex Method for Quadratic Programming," *Econometrica*, **27**, pp. 382–398, 1959.
13. M. Frank and P. Wolfe, "An Algorithm for Quadratic Programming," *Naval Res. Logistics Quart.*, **3**, pp. 95–110, 1956.
14. J. D. Pearson, "On Variable Metric Methods of Minimization," *Computer J.*, **11**, pp. 171–178, 1969.
15. G. Zoutendijk, "Nonlinear Programming: A Numerical Survey," *J. SIAM Control*, **4**, pp. 194–210, 1966.
16. G. Zoutendijk, *Methods of Feasible Directions*, American Elsevier, New York, 1960.
17. J. B. Rosen, "The Gradient Projection Method for Nonlinear Programming, Part I, Linear Constraints," *J. SIAM*, **8**, pp. 181–217, 1960.
18. J. B. Rosen, "The Gradient Projection Method for Nonlinear Programming, Part II, Nonlinear Constraints," *J. SIAM*, **9**, pp. 514–532, 1961.
19. C. W. Carroll, "The Created Response Surface Technique for Optimizing Nonlinear Restrained Systems," *Operations Res.*, **9**, pp. 169–184, 1961.
20. A. V. Fiacco and G. P. McCormick, *Nonlinear Programming: Sequential Unconstrained Minimization Techniques*, Wiley, New York, 1968.
21. D. J. Wilde, *Optimum Seeking Methods*, Prentice-Hall, Englewood Cliffs, N.J., 1964.
22. D. J. Wilde and C. S. Beightler, *Foundation of Optimization*, Prentice-Hall, Englewood Cliffs, N.J., 1967.
23. G. P. McCormick and J. D. Pearson, "Variable Metric Methods and Unconstrained Optimization," Joint Conference on Optimization, University of Keele, Staffordshire, England, March 1968.
24. G. E. Myers, "Properties of the Conjugate Gradient and Davidon Methods," *J. Optimization Theory Applications*, **2**, pp. 209–219, 1968.

25. D. Tabak, "Comparative Study of Various Minimization Techniques Used in Mathematical Programming," *IEEE Trans. Automatic Control*, **AC-14**, p. 572, 1969.

26. W. C. Davidon, "Variable Metric Method of Minimization," *Report ANL-5990* (rev), Argonne National Laboratory, Argonne, Ill., 1959.

27. R. Fletcher and M. J. D. Powell, "A Rapidly Convergent Descent Method for Minimization," *Computer J.*, **6**, pp. 163–168, 1963.

28. M. R. Hestenes, "The Conjugate Gradient Method for Solving Linear Systems," *Proceedings of the Symposium on Applied Mathematics*, vol. VI, McGraw-Hill, New York, 1956, pp. 83–102.

29. H. J. Kelley and G. E. Myers, "Conjugate Direction Methods for Parameter Optimization," *Astronautica Acta* (to appear).

30. L. S. Lasdon, S. K. Mitter, and A. D. Waren, "The Conjugate Gradient Method for Optimal Control Problems," *IEEE Trans. Automatic Control*, **AC-12**, pp. 132–138, April 1967.

31. S. S. Tripathi and K. S. Narendra, "Conjugate Direction Methods for Nonlinear Optimization Problems," National Electronics Conference, Chicago, Dec. 1968.

4

OPTIMAL CONTROL AND MATHEMATICAL PROGRAMMING

4.1 INTRODUCTION

The basic notions of mathematical programming have been discussed in Chapter 2, while the computational algorithms for numerical solution of mathematical-programming problems have been reviewed in Chapter 3. Considering the mathematical-programming techniques along with optimal-control problems, several questions may arise:

1. What is the connection between a mathematical-programming and an optimal-control problem?
2. How can we utilize the computational algorithms of mathematical programming to provide efficient numerical solutions of optimal-control problems?

It is the purpose of this chapter to designate the initial steps in attempting to answer these questions. There exists a vast amount of literature concerning various aspects of optimal control. Only a few chosen representatives have been referred to at the end of the chapter [1–3]. To provide a common denominator for the discussion of optimal-control problems, a specific form of such a problem is formulated in Sec. 4.2. This formulation will serve as a basis for the discussion in the remainder of this chapter as well as in the following chapters. An initial answer to question 1 will be attempted in Sec. 4.3. Without entering into further details, we may look at this question as follows:

Both mathematical-programming and optimal-control problems are optimization problems. In other words, they are multivariable extremization problems, subject to various side conditions, or constraints. Naturally, we can expect that there should be a definite connection between them. This is indeed demonstrated in Sec. 4.3.

Question 2, or rather only a few aspects of it, is tackled in Sec. 4.4. Some general considerations arising in connection with applying mathematical programming to the numerical solution of optimal-control problems are discussed. This subject is continued in much more detail in the subsequent chapters, along with illustrative computational examples and description of various technical applications.

4.2 FORMULATION OF THE PROBLEM

First, an optimal control problem [1–3] for a continuous-time dynamic system will be considered. Various particular cases will be treated in subsequent chapters. The system under consideration is assumed to be nonautonomous, to be nonlinear, and to have lumped parameters. The dynamic behavior of the system may be described by a set of ordinary, nonlinear, first-order differential equations

$$\dot{\mathbf{y}} = \mathbf{f}(\mathbf{y}, \mathbf{u}, t) \qquad (4\text{-}1)$$

where $\mathbf{y} = n$-dimensional state vector $= [y_1, y_2, \ldots, y_n]^T$
$\mathbf{u} = m$-dimensional control vector $= [u_1, u_2, \ldots, u_m]^T$
$t = $ Time
$\mathbf{f} = n$-dimensional nonlinear vector function
$\dot{\mathbf{y}} = d\mathbf{y}/dt$

It is assumed that at a certain initial time t_0, the system is in a given state

$$\mathbf{y}(t_0) = \mathbf{y}_0 \qquad (4\text{-}2)$$

This assumption is valid in most practical applications. However, there may be cases where the initial state is an unknown variable. If this is the case, some intermediate state or the final state would usually be specified, which would not change the basic approach to the problem, since we can always solve the problem using the reversed-time approach. In some cases, the state vector may not be prespecified at any time.

For the control system just described, the objective of the control action is to bring the system from its initial state to a new state which is inside the region defined by

$$s(\mathbf{y}_f) \geqslant 0 \tag{4-3}$$

at time $t = t_f$, where $\mathbf{s}$ is a k-dimensional nonlinear vector function which defines the *terminal region*. The final state, generally free, is denoted as

$$\mathbf{y}_f = \mathbf{y}(t_f) \tag{4-4}$$

Equation (4-3), in general, defines a closed region bounded by the surfaces

$$s_i(\mathbf{y}_f) = 0, \quad i = 1, \ldots, k \tag{4-5}$$

The final state $\mathbf{y}_f$ may lie either inside the region described by Eq. (4-3) or on its boundary, which is defined by Eq. (4-5). In particular, the terminal region may consist of just one fixed state $\mathbf{y}_f$. The final time, t_f, may be either fixed or free.

The control action of bringing the state of the system from $\mathbf{y}_0$ to $\mathbf{y}_f$ should be performed in a fashion such that a specified performance index is extremized. A general form of the performance index may be chosen as

$$J = \int_{t_0}^{t_f} F[\mathbf{y}(t), \mathbf{u}(t), t] \, dt + H(\mathbf{y}_f) \tag{4-6}$$

where F and H are nonlinear functions of the state and control variables. As shown by Eq. (4-6), the performance index may depend either on the trajectory along which the control action is performed on the terminal state, or on both. Of particular interest is the time-optimal problem for which $F = 1$ and $H = 0$; that is,

$$J = \int_{t_0}^{t_f} dt = (t_f - t_0) \tag{4-7}$$

where t_f is free.

In most practical applications, additional constraints may be imposed on the state and control variables. These constraints can usually be expressed as a set of inequalities or equalities of the following form:

$$\mathbf{g}(\mathbf{y}, \mathbf{u}) \geqslant 0 \tag{4-8}$$

$$\mathbf{h}(\mathbf{y}, \mathbf{u}) = 0 \tag{4-9}$$

where $\mathbf{g}$ and $\mathbf{h}$ are, in general, nonlinear vector functions. For instance, one very often finds that the absolute values of the state and control variables are limited so that the following inequality constraints are imposed at all times:

$$\begin{aligned} u_{i\,\max} - |u_i| \geqslant 0, & \quad i = 1, \ldots, m \\ y_{i\,\max} - |y_i| \geqslant 0, & \quad i = 1, \ldots, n \end{aligned} \tag{4-10}$$

where $u_{i\,\max}$ and $y_{i\,\max}$ are specified scalar values. Equations (4-10) are, of course, some of the simplest forms of constraints that may be imposed on the state and control variables.

The continuous-time optimal-control problem may now be stated as follows:

Find an optimal control vector $\mathbf{u}(t)$, $t_0 \leqslant t \leqslant t_f$, which transfers the system from the initial state $\mathbf{y}_0$ to the terminal region $\mathbf{s}$, extremizing the performance index in Eq. (4-6) and satisfying the system dynamic equations in Eq. (4-1), the terminal conditions in Eq. (4-3), and the constraints in Eqs. (4-8) and (4-9).

Or, we may write

$$\min \text{ (or max)} \left\{ \int_{t_0}^{t_f} F(\mathbf{y}, \mathbf{u}, t)\, dt + H(\mathbf{y}_f) \,\Big|\, \dot{\mathbf{y}} = \mathbf{f}(\mathbf{y}, \mathbf{u}, t); \right.$$
$$\left. \mathbf{y}(t_0) = \mathbf{y}_0;\ \mathbf{s}(\mathbf{y}_f) \geqslant \mathbf{0};\ \mathbf{g}(\mathbf{y}, \mathbf{u}) \geqslant \mathbf{0};\ \mathbf{h}(\mathbf{y}, \mathbf{u}) = \mathbf{0} \right\} \quad (4\text{-}11)$$

In what follows, the connection between the optimal-control problem defined by Eq. (4-11) and the mathematical-programming problem presented in Chapter 2 will be discussed.

4.3 MATHEMATICAL PROGRAMMING AND OPTIMAL CONTROL

The connection between the mathematical-programming problem, presented in Chapter 2, and the optimal-control problem, presented in Sec. 4.2, is quite obvious. An optimal-control problem is actually a mathematical-programming problem. This fact has been noted by several authors, for instance, in reference 4, p. XV. To see this more clearly, we have to represent the optimal-control problem in Eq. (4-11) in a somewhat different way.

First, let us observe the integral part of the performance index in Eq. (4-6).

An integral may be represented as a limit of a summation [5]. To accomplish this for the integral in Eq. (4-6), we have to partition the time interval from t_0 to t_f into n subintervals $(t_1 - t_0), (t_2 - t_1), \ldots, (t_f - t_{n-1})$, where $t_f = t_n$. Within each interval $t_k - t_{k-1}$ ($k = 1, \ldots, n$), we may pick an arbitrary t'_k, with $t_{k-1} \leqslant t'_k < t_k$. Using this partition, the integral in Eq. (4-6) is written

$$\int_{t_0}^{t_f} F[\mathbf{y}(t), \mathbf{u}(t), t]\, dt = \lim_{n \to \infty} \sum_{k=1}^{n} F[\mathbf{y}(t'_k), \mathbf{u}(t'_k), t'_k](t_k - t_{k-1}) \quad (4\text{-}12)$$

The final time t_f is, of course, finite, but the total number of time subintervals n goes to infinity. For the same time partition, the state equations in Eq. (4-1) can be rewritten as

$$\lim_{h_k \to 0} \frac{1}{h_k}[\mathbf{y}(t_k) - \mathbf{y}(t_{k-1})] = \mathbf{f}[\mathbf{y}(t_{k-1}), \mathbf{u}(t_{k-1}), t_{k-1}] \quad (4\text{-}13)$$

where

$$h_k = t_k - t_{k-1}, \quad k = 1, \ldots, n; \quad n \to \infty \quad (4\text{-}14)$$

Similarly, we may rewrite the additional constraints in Eqs. (4-8) and (4-9) as

$$\mathbf{g}[\mathbf{y}(t_k), \mathbf{u}(t_k)] \geqslant 0$$
$$\mathbf{h}[\mathbf{y}(t_k), \mathbf{u}(t_k)] = 0, \quad k = 0, 1, \ldots, n; \quad n \to \infty \quad (4\text{-}15)$$

The terminal conditions in Eq. (4-3) and the nonintegral term of the performance index in Eq. (4-6), $H(\mathbf{y}_f)$, are functions of one state point only and therefore remain unchanged. We can now combine Eqs. (4-12) through (4-15) into a formulation of a mathematical-programming problem, similar to the one in Eq. (4-11):

$$\min \text{ (or max)} \left\{ \lim_{n \to \infty} \sum_{k=1}^{n} F[\mathbf{y}(t'_k), \mathbf{u}(t'_k), t'_k](t_k - t_{k-1}) \right.$$

$$+ H[\mathbf{y}(t_f)] \Big| \lim_{h_k \to 0} \frac{1}{h_k}[\mathbf{y}(t_k) - \mathbf{y}(t_{k-1})] = \mathbf{f}[\mathbf{y}(t_{k-1}), \mathbf{u}(t_{k-1}), t_{k-1}],$$

$$h_k = t_k - t_{k-1}, k = 1, \ldots, n; n \to \infty; \mathbf{s}[\mathbf{y}(t_f)] \geqslant 0;$$

$$\left. \mathbf{g}[\mathbf{y}(t_k), \mathbf{u}(t_k)] \geqslant 0, \mathbf{h}[\mathbf{y}(t_k), \mathbf{u}(t_k)] = 0, k = 0, \ldots, n; n \to \infty \right\} \quad (4\text{-}16)$$

As we can see, the problem formulated in Eq. (4-16) is indeed a mathematical-programming problem, whose unknown variables are

$$\mathbf{y}(t_k), \mathbf{y}(t'_k), k = 1, \ldots, n; \quad \mathbf{u}(t_k), \mathbf{u}(t'_k), k = 0, 1, \ldots, n-1; \quad n \to \infty$$

The difference between this MP problem and the one discussed in Chapter 2 is that there is an infinite number of unknown variables in the problem in Eq. (4-16), while the number of variables in the usual MP problem is finite. In other words, the continuous-time *optimal-control problem* constitutes a mathematical-programming problem of infinite dimension, i.e., *a mathematical-programming problem in an infinite dimensional space.*

The extension of Kuhn-Tucker's theorem to problems in infinite dimensional spaces was originally worked out by Hurwicz [6]. Subsequently, the idea was extended to optimal-control theory by other authors [7–11]. In the works cited, necessary and sufficient conditions for the solution of optimal-control problems have been proved and connection with the Kuhn-Tucker-Hurwicz conditions established.

In view of the fact that the optimal-control problem is indeed a mathematical-programming problem, there should exist a direct connection between the commonly accepted methods of treating optimal-control problems and the methods formally used in mathematical programming. One of the most popular methods in optimal control is the maximum principle of Pontryagin [12]. Halkin and Neustadt [13, 14] have derived a general maximum principle which provides necessary conditions for the solution of general optimization problems. They have shown that Kuhn-Tucker conditions, both for finite and infinite spaces, may be derived from their conditions. Similar work has been done by other authors. For instance, Hanson [15] has shown that certain forms of MP problems have their analogous counterparts in optimal control. Ho and Brentani [16], using a differential approach, have developed a computational method for treating optimal-control problems with inequality constraints. They have proved the sufficiency of their conditions using Kuhn-Tucker's saddle-point conditions, thereby demonstrating the connection between an optimal-control problem and MP methods.

An elegant way of presenting the different optimization techniques in a unified manner was recently proposed by Wilde and Beightler [17]. These authors have accomplished the unification by introducing some special concepts, which are described as follows: An optimization problem involving n variables and m ($m < n$) equality constraints was considered. Out of the n variables, m were chosen and defined as "state variables." The state variables were expressed as functions of the remaining $n - m$ variables, which were defined as "decision variables." An important concept introduced was that of the *constrained derivative* of the performance index with respect to the decision variables within the feasible region. Similarly, *sensitivity coefficients*, which are partial derivatives of the optimal performance index with respect to perturbations in the equality constraints, were defined. Next, the concept of *slack derivatives* was introduced. The slack derivatives are conceptually the same as the sensitivity coefficients. The difference between them is that the slack derivatives are defined in every point of the feasible region, while the sensitivity coefficients are defined in the neighborhood of the extremum only. Using these notions, the authors have rederived both the Kuhn-Tucker conditions and the maximum principle, thereby demonstrating the close connection between them.

4.4 COMPUTATIONAL CONSIDERATIONS

It is well known that a numerical solution of continuous differential equations involves discretizing the equations with respect to the independent variables. The same applies to the continuous optimal-control problem. No matter what method or technique is used, the differential equations involved, as well as the integral of the performance index, are to be discretized in time in order to obtain a computerized solution. In other words, the differential equations are replaced by difference equations, and the integral by a finite summation. The mathematical-programming optimal-control problem, when it comes to the actual computation, may have quite an enormous number of variables. However, this number is finite. In many cases this number may be reduced to the extent that actual implementation becomes practical. Cases such as this will be discussed in subsequent chapters.

The optimal-control problem formulated in Sec. 4.2 will now be reformulated in discretized form [18]. The entire time interval between the initial time t_0 and the final time t_f is subdivided into N generally unequal time intervals. The discrete time instants considered are designated as

$$t_0, t_1, t_2, \ldots, t_{N-1}, t_N = t_f$$

and the time intervals between them are

$$T_1, T_2, \ldots, T_N$$

Therefore,

$$T_i = t_i - t_{i-1} \qquad (4\text{-}17)$$

and

$$\sum_{i=1}^{N} T_i = T = t_f - t_0 \qquad (4\text{-}18)$$

Usually, T, as well as all the T_i's, are unknown a priori and have to be found as part of the solution of the optimization problem. This explicit representation of the time intervals as separate problem variables permits the imposition of explicit constraints on them.

The integral part of the performance index of Eq. (4-6) is rewritten as a summation,

$$J = \sum_{i=1}^{N} F[\mathbf{y}(i), \mathbf{u}(i-1), t_i] T_i + H[\mathbf{y}(N)] \qquad (4\text{-}19)$$

where

$$\mathbf{y}(i) = \mathbf{y}(t_i)$$
$$\mathbf{u}(i) = \mathbf{u}(t_i)$$

This can be regarded as the simplest form of an approximation to Eq. (4-6). There are, of course, much more elaborate forms of numerical integration [19]. In many cases the performance index may be initially defined as a summation,

$$J = \sum_{i=1}^{N} F[\mathbf{y}(i), \mathbf{u}(i-1), t_i] + H[\mathbf{y}(N)] \qquad (4\text{-}20)$$

In formulating the optimal-control problem as a mathematical-programming problem, each scalar component of the state or control vector is considered as a separate variable at each discrete time t_i ($i = 1, \ldots, N$). As mentioned before, the time intervals T_i ($i = 1, \ldots, N$) are considered as problem variables as well. All the variables of the MP problem are considered as components of a variables vector $\mathbf{x}$. Each variable is denoted by $x(i)$ ($i = 1, \ldots, N_v$), where N_v is the total number of variables considered in the MP problem. The scalar components of the control problem can be assigned as the $x(i)$ variables in the following way. The order of assignment is, of course, irrelevant.

Control-problem variable	Mathematical-programming variables
T_1	$x(1)$
T_2	$x(2)$
.	.
.	.
.	.
T_N	$x(N)$
$u_1(0)$	$x(N+1)$
$u_2(0)$	$x(N+2)$
.	.
.	.
.	.
$u_m(0)$	$x(N+m)$
$u_1(1)$	$x(N+m+1)$
$u_2(1)$	$x(N+m+2)$
.	.
.	.
.	.
$u_m(1)$	$x(N+2m)$
.	.
.	.
.	.
$u_1(N-1)$	$x[N+(N-1)m+1]$
.	.
.	.
.	.
$u_m(N-1)$	$x(N+Nm)$
$y_1(1)$	$x(N+Nm+1)$
$y_2(1)$	$x(N+Nm+2)$
.	.
.	.
.	.
$y_n(1)$	$x(N+Nm+n)$
.	.
.	.
.	.
$y_1(N)$	$x[N+Nm+(N-1)n+1]$
$y_2(N)$	$x[N+Nm+(N-1)n+2]$
.	.
.	.
.	.
$y_n(N)$	$x(N+Nm+Nn)$

This gives a total of $N_v = N(n+m+1)$ variables.

The notation for the control and state components is, for instance, $u_j(i)$ = the jth component of **u** at $t = t_i$.

For the MP variables, $x(i)$ is simply the ith variable of the problem, or the ith component of vector **x**.

The variables $x(i)$ ($i = 1, \ldots, N_v$) are now substituted into the performance index of Eq. (4-19) or (4-20), into the terminal constraint in Eq. (4-3), as well as into all other constraints, such as in Eqs. (4-8) and (4-9) for the control, state, and time variables, whichever are applicable.

The state equations of the system are, of course, to be satisfied at any time, and the treatment of these presents a particular difficulty when applying mathematical programming.

Basically there are two different cases in the treatment of the state equations:

1. The closed-form solution of the state equations in Eq. (4-1) is known.
2. The closed-form solution of Eq. (4-1) is unknown.

The two cases will be discussed in the following.

1. Closed-Form Solution of the State Equation Is Known

The solution of Eq. (4-1) at any time t can be written as

$$\mathbf{y}(t) = \boldsymbol{\phi}[t, t_0, \mathbf{u}(t_0), \mathbf{y}(t_0)] \tag{4-21}$$

where $\boldsymbol{\phi}$ us ab n-dimensional vector function. Between two consecutive time instants t_{i-1} and t_i, Eq. (4-21) may be written as

$$\mathbf{y}(i) - \boldsymbol{\phi}[T_i, \mathbf{u}(i-1), \mathbf{y}(i-1)] = \mathbf{0} \tag{4-22}$$

Expressing Eq. (4-22) in component form ($i = 1, \ldots, N$), one has nN equality constraints in the form of Eq. (4-9) for the MP problem.

Equation (4-22) can also be regarded as the difference state equations of a basically discrete-time system.

The control vector $\mathbf{u}(i-1)$ is assumed constant during the period $t_{i-1} \leqslant t \leqslant t_i$. This is, of course, an approximation for continuous systems. However, it is the exact description for sampled-data systems with zero-order hold.

In applying mathematical programming, the $x(i)$ variables are substituted in Eq. (4-22) for the $\mathbf{y}$, $\mathbf{u}$, and T_i variables, exactly as in all other constraints.

2. Closed-Form Solution Is Unknown

In this case, we have to resort to what is usually done when a closed-form solution of a differential equation is unknown, namely, numerical integration. In other words, the state equations in Eq. (4-1) are discretized and written as difference equations:

$$\mathbf{y}(i+1) - \mathbf{y}(i) = T_{i+1}\mathbf{f}[\mathbf{y}(i), \mathbf{u}(i), t_i], \quad i = 0, 1, \ldots, N-1 \tag{4-23}$$

Equation (4-23) may now serve as nN equality constraints, in the same manner as Eq. (4-22). It should be realized that while Eq. (4-22) is exact, Eq. (4-23) is an approximation, and the results obtained after using it should be interpreted accordingly. A theoretical justification of some discretization schemes for continuous-time optimal control problems was recently given by J. Cullum [20].

If we have a system with very rapidly changing state variables, i.e., a high-frequency oscillatory system, this approximation might not work. On the other hand, if the time change of the state variables is relatively slow, we may obtain a desired degree of precision by limiting the discrete-time intervals from above:

$$T_i \leqslant T_{max}, \quad i = 1, \ldots, N \tag{4-24}$$

In some sampled-data systems, we may have, in addition, a minimum-sampling-period restriction, for practical reasons:

$$T_i \geqslant T_{min}, \quad i = 1, \ldots, N \tag{4-25}$$

We should also note that Eq. (4-23) is obtained using a very simple difference scheme. With more elaborate difference schemes [19], the accuracy of the solutions may be improved.

To summarize, let us again express the optimal-control problem in a condensed mathematical-programming formulation. Since it is the case in the majority of problems, it will be assumed that the closed-form solution of the state equations is unknown.

$$\min \text{ (or max)} \left\{ \sum_{i=1}^{N} F[\mathbf{y}(i), \mathbf{u}(i-1), t_i] + H[\mathbf{y}(N)] \,|\, \mathbf{s}[\mathbf{y}(N)] \geqslant \mathbf{0};\right.$$

$$\mathbf{g}(\mathbf{y}, \mathbf{u}, T_i) \geqslant \mathbf{0}; \quad \mathbf{h}(\mathbf{y}, \mathbf{u}, T_i) = \mathbf{0}; \quad i = 1, \ldots, N;$$

$$\left. \mathbf{y}(i+1) - \mathbf{y}(i) = T_{i+1} \mathbf{f}[\mathbf{y}(i), \mathbf{u}(i), t_i]; i = 0, 1, \ldots, N-1 \right\} \tag{4-26}$$

As we can see, the functions **g** and **h** were formulated as dependent on the sampling intervals T_i ($i = 1, \ldots, N$) to provide the option of imposing explicit constraints on the time intervals. Subsequent chapters will be devoted to the discussion of various particular cases of the problem in Eq. (4-26).

REFERENCES

1. A. P. Sage, *Optimum Systems Control*, Prentice-Hall, Englewood Cliffs, N.J., 1968.
2. M. Athans and P. L. Falb, *Optimal Control*, McGraw-Hill, New York, 1966.
3. E. B. Lee and L. Markus, *Foundations of Optimal Control Theory*, Wiley, New York, 1967.
4. J. Abadie, ed., *Nonlinear Programming*, Wiley, New York, 1967.
5. R. Courant, *Differential and Integral Calculus*, vol. 1, 2nd ed., Wiley-Interscience, New York, 1957.
6. L. Hurwicz, "Programming in Linear Spaces," in *Studies in Linear and Non-*

linear Programming, by K. J. Arrow, L. Hurwicz, and H. Uzawa, Stanford University Press, Stanford, Cal., 1958, pp. 38–102.

7. D. L. Russell, "The Kuhn-Tucker Conditions in Banach Space with an Application to Control Theory," *J. Math. Anal. Appl.*, **15**, pp. 200–212, 1966.

8. B. N. Pshenichniy, "Linear Optimal Control Problems," *J. SIAM Control*, **4**, pp. 577–593, 1966.

9. R. M. Van Slyke, "A Duality Theory for Abstract Mathematical Programs with Applications to Optimal Control Theory," *Report D1-82-0671*, Boeing Scientific Research Laboratories, Seattle, Wash., Oct. 1967.

10. J. Ponstein, "Multiplier Functions in Optimal Control," *J. SIAM Control*, **6**, pp. 648–658, 1968.

11. D. O. Norris, "Lagrangian Saddle Points and Optimal Control," *J. SIAM Control*, **5**, pp. 594–599, 1967.

12. L. S. Pontryagin, V. G. Boltianskii, R. V. Gamkrelidze, and E. F. Mishchenko, *The Mathematical Theory of Optimal Processes*, Wiley-Interscience, New York, 1962.

13. H. Halkin and L. W. Neustadt, "General Necessary Conditions for Optimization Problems," *Proc. Natl. Acad, Sci. U.S.*, **56**, pp. 1066–1071, 1966.

14. L. W. Neustadt, "An Abstract Variational Theory with Applications to a Broad Class of Optimization Problems," *J. SIAM Control*, part I: General Theory, **4**, pp. 505–527, 1966; part II: Applications, **5**, pp. 90–137, 1967.

15. M. A. Hanson, "Bounds for Functionally Convex Optimal Control Problems," *J. Math. Anal. Appl.*, **8**, pp. 84–89, 1964.

16. Y. C. Ho and P. B. Brentani, "On Computing Optimal Control with Ineqality Constraints," *J. SIAM Control*, **1**, pp. 319–348, 1963.

17. D. J. Wilde and C. S. Beightler, *Foundations of Optimization*, Prentice-Hall, Englewood Cliffs, N.J., 1967.

18. D. Tabak and B. C. Kuo, "Application of Mathematical Programming in the Design of Optimal Control Systems," *Intern. J. Control*, **10**, pp. 545–552, 1969.

19. J. B. Scarborough, *Numerical Mathematical Analysis*, 5th ed., The Johns Hopkins Press, Baltimore, 1963.

20. J. Cullum, "Discrete Approximations to Continuous Optimal Control Problems," *SIAM J. Control*, **7**, pp. 32–49, 1969.

5

LINEAR CONTINUOUS-TIME SYSTEMS

5.1 INTRODUCTION

As in many other areas of technology, the solution of optimal control problems for linear systems is analytically simpler to handle than that of nonlinear systems. One of the main reasons for this is that the closed-form solution of a set of linear state equations is known, which facilitates the analytical treatment of problems of this kind. It is partly for this reason that most of the work done in the area of solving optimal-control problems by mathematical programming was concentrated on linear systems. Another important reason for preferring the treatment of linear systems is that in many cases one can reduce an optimal-control problem for a linear system to a linear-programming problem, which is much easier to handle computationally. Actually, as mentioned in Chapter 3, it is the only class of MP in which there is a guaranteed global extremum in a finite number of iterations. Even if the extremum is infinite, or if there is no solution, the existing algorithms would readily provide this information. This, of course, does not hold for general nonlinear-programming problems, with the exception of the class of quadratic-programming problems discussed in Chapter 3. Therefore, it is always recommended that an effort be made to formulate the optimal-control problem as an LP or QP problem whenever possible.

In this chapter, several methods of formulating an optimal-control problem for linear, continuous-time systems as a MP problem will be discussed. Examples illustrating the various methods of the mentioned formulation, with stress on technical applications, are included.

5.2 FORMULATION OF THE PROBLEM

The optimal-control problem discussed in Chapter 4 will now be reformulated for a linear control system—in other words, for a system whose dynamic state equations are linear. Following the notation used in Chapter 4, we can write the set of state equations for a linear nth order system as

$$\dot{\mathbf{y}} = A\mathbf{y} + B\mathbf{u} \tag{5-1}$$

where $\mathbf{y} = n$-dimensional state vector
$\mathbf{u} = m$-dimensional control vector
$A = n \times n$ constant matrix
$B = n \times m$ constant matrix

In general, the matrices A and B are time dependent. However, in this chapter, they will be assumed to be constant. In the case when A and B are time varying, we have the following solution of the state equation (5-1) [9]:

$$\mathbf{y}(t) = \Phi(t, t_0)\mathbf{y}(t_0) + \int_{t_0}^{t} \Phi(t, \tau)B(\tau)\mathbf{u}(\tau)\,d\tau$$

where

$$\frac{\partial \Phi(t, \tau)}{\partial t} = A(t)\Phi(t, \tau), \qquad \Phi(t, t) = I$$

In general, we can obtain the explicit solution of the time-varying state equations through numerical integration. A purely analytic solution as in the case of time-invariant systems [9] is generally unknown. Therefore, this type of problem is classified under nonlinear systems, and the subject will be treated in Chapter 6.

The performance index for a linear system can be expressed in the same general way as in Eq. (4-6), (4-19), or (4-20). Of particular interest, however, are performance indices expressible in linear or quadratic form for the obvious reasons stated in Sec. 5.1.

Some of the most popularly used performance indices are as follows [1–8]:

1. Minimum energy problem:

$$J = \int_{t_0}^{t_f} u^2(t)\,dt \tag{5-2}$$

2. Minimum fuel problem:

$$J = \int_{t_0}^{t_f} |u(t)| \, dt \tag{5-3}$$

3. Minimum time problem:

$$J = \int_{t_0}^{t_f} dt = (t_f - t_0) \tag{5-4}$$

These problems will be treated in more detail in the following sections of this chapter. It should be stressed that this list of possible performance indices is by no means exhaustive. Additional ones may be considered depending on the particular problem.

In addition to the performance index, constraints of the form given in Eqs. (4-8) and (4-9) can be imposed. If any of the **g** or **h** functions are nonlinear, we would eventually wind up with a NLP problem, even though the control system itself is linear. In many applications, it is sufficient to pose linear constraints of the type in Eqs. (4-10) and (4-11).

The closed-form solution of the state equations in Eq. (5-1) can be expressed at any time t:

$$\mathbf{y}(t) = \Phi(t)\left[\mathbf{y}_0 + \int_0^t \Phi^{-1}(\tau)B\mathbf{u}(\tau)\,d\tau\right] \tag{5-5}$$

where $\Phi(t)$ is the $n \times n$ state transition matrix, and $\mathbf{y}_0$ is the initial state at $t = 0$.

Using the time discretization procedure as in sec. 4.4, we can rewrite Eq. (5-5) between two consecutive discrete-time instants t_{j-1} and t_j as

$$\mathbf{y}(j) = \Phi(t_j, t_{j-1})\left[\mathbf{y}(j-1) + \int_{t_{j-1}}^{t_j} \Phi(t_{j-1}, \tau)B\,d\tau\,\mathbf{u}(j-1)\right], \tag{5-6}$$
$$j = 1, \ldots, N$$

For linear time-invariant systems, the state transition matrix $\Phi(t)$ has the form [9]

$$\Phi(t) = e^{At} \tag{5-7}$$

and

$$\Phi(t_j, t_{j-1}) = \Phi(t_j - t_{j-1}) = e^{A(t_j - t_{j-1})} = e^{AT_j} \tag{5-8}$$

The discretized state transition equations of Eq. (5-6) can now be rewritten as a set of equality constraints in the following form:

$$\mathbf{y}(j) - e^{AT_j}\left[\mathbf{y}(j-1) + \int_{t_{j-1}}^{t_j} e^{A(t_j - \tau)}B\,d\tau\,\mathbf{u}(j-1)\right] = 0, \tag{5-9}$$
$$j = 1, \ldots, N$$

Equation (5-6) or (5-9) constitutes a set of nN equality constraint equations. (n is the dimension of **y**.)

It should be stressed that in the formulation of Eqs. (5-6) and (5-9) the control vector $\mathbf{u}(j-1)$ is assumed *constant* in the time interval $t_{j-1} \leqslant t \leqslant t_j$. This assumption is necessary in order to obtain a numerical solution in the more general cases of linear systems. As will be explained later, we can avoid this assumption for a certain class of systems by use of the so-called generalized programming [7].

To summarize, a typical MP formulation of an optimal-control problem, for a linear system, is as follows:

$$\min \left\{ \sum_{i=1}^{N} F[\mathbf{y}(i), \mathbf{u}(i-1), i] \,\bigg|\, u_{i\,\max} - |u_i| \geqslant 0, i = 1, \ldots, m; \right.$$
$$y_{i\,\max} - |y_i| \geqslant 0, i = 1, \ldots, n; \mathbf{s}[y(N)] \geqslant \mathbf{0}; \quad (5\text{-}10)$$
$$\left. \mathbf{y}(j) - e^{AT_j}\left[\mathbf{y}(j-1) + \int_{t_{j-1}}^{t_j} e^{A(t_{j-1}-\tau)} B\, d\tau\, \mathbf{u}(j-1)\right] = \mathbf{0}, j = 1, \ldots, N \right\}$$

where the original, integral form performance index is expressed in discrete-time form.

This, of course, is not the most general and exhaustive formulation. However, it encompasses a wide class of applications. The performance index can take any of the forms given in Eqs. (5-2) through (5-5), or any other form. The final region described by the set of inequality constraints, $\mathbf{s} \geqslant \mathbf{0}$, may occasionally consist of only one point. Usually, the initial state $\mathbf{y}(0)$ is specified, but sometimes it may be defined as a set $\mathbf{y}(0) \in Y_0$. The functions $\mathbf{s}$ which define the terminal region may, in general, be nonlinear. The state equations equality constraints are also nonlinear in the most general case, i.e., when the sampling periods T_j are not specified and are part of the unknown problem variables.

To summarize the argument, we can say that dealing with a linear system does not necessarily imply that we would have strictly linear constraints in the MP formulation of the problem. Some of the constraints may be nonlinear in more general cases. On the other hand, to obtain a MP problem with strictly linear constraints, we have to deal with a linear or linearized system, i.e., a system whose state equations are linear. Observing Eq. (5-10), we can add that in order to achieve linearity of all of the constraints in the problem under consideration in Eq. (5-10), the following conditions are to be satisfied:

1. The sampling periods T_j ($j = 1, \ldots, N$) are to be specified a priori; however, they may, in general, be unequal.
2. The final set functions $\mathbf{s}[\mathbf{y}(N)]$ are to be linear.

To obtain a LP problem, the F function in Eq. (5-10) has to be linear with respect to all its variables, and to obtain a QP problem, F has to be quadratic. In the subsequent sections, examples dealing with various possibilities of problem formulation will be presented.

5.3 A LINEAR SYSTEM WITH UNSPECIFIED SAMPLING INTERVALS

An example concerning the more general case of a MP formulation of an optimal-control problem will be presented in this section [5, 10]. The generality of the case depends on the fact that the discretized time intervals used in the MP formulation are unknown a priori. The time intervals are actually part of the variables of the optimal-control and MP problem whose values are sought as the end product of the solution. Inspecting Eq. (5-10), we can see that the state difference equations are nonlinear because of the e^{AT_j} and $e^{A(t_j-\tau)}$ terms, since T_j and t_j are unknown a priori. These equations will form a set of nN nonlinear equality constraints in the MP problem.

To illustrate the method described above, an example of a linear system will be given. The state equations of the system considered are

$$\dot{y}_1 = y_2$$
$$\dot{y}_2 = -.16y_1 - .032y_2 + .16u \tag{5-11}$$

The purpose of the control operation is to drive the system from the initial state

$$\mathbf{y}(0) = \begin{bmatrix} 0 \\ 0 \end{bmatrix}$$

to the final state

$$\mathbf{y}_f = \begin{bmatrix} y_{1f} \\ y_{2f} \end{bmatrix}$$

so that a certain quadratic performance index is minimized and so that no overshoot occurs before the target point is reached. In this problem, the value of the control input u is assumed known; only the time periods, after which this value is changed, are treated as unknown variables.

Suppose that the desired final state is given as

$$\mathbf{y}_f = \begin{bmatrix} 2 \\ 0 \end{bmatrix}$$

and let us set the control variable u as shown in Fig. 5-1. The final time is denoted as $t_f = t_4$, and the total number of periods is chosen to be 4.

The times $t_1, \ldots, t_4$ or the periods $T_1, \ldots, T_4$ are to be determined as a result of the optimization computation. The decrease in value of u at t_2 is intended mainly to prevent the overshoot of y_1. At the final time t_4, y_1 reaches the desired value of $y_{1f} = 2$ or a specified percentage of it. As we can see, in a problem such as this, the total number of periods considered is

$$N = y_{1f} + 2$$

The state transition equations of Eq. (5-11) are written as

$$\mathbf{y}(t) = \Phi(t, t_0)\mathbf{y}(t_0) + \mathbf{b}(t, t_0)u(t_0) \tag{5-12}$$

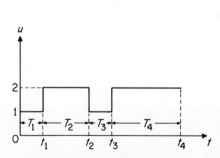

FIG. 5-1. The control variable.

where

$$\Phi(t, t_0) = e^{-.016(t-t_0)} \begin{bmatrix} \sin[.4(t-t_0) + 87.7°] & 2.5\sin.4(t-t_0) \\ -.4\sin.4(t-t_0) & -\sin[.4(t-t_0) - 87.7°] \end{bmatrix}$$

and

$$\mathbf{b} = \begin{bmatrix} 1 - e^{-.016(t-t_0)}\sin[.4(t-t_0) + 87.7°] \\ .4e^{-.016(t-t_0)}\sin.4(t-t_0) \end{bmatrix}$$

During the time interval T_j, Eq. (5-12) can be written as

$$y_1(j) = e^{-.016T_j}[\sin(.4T_j + 1.531)y_1(j-1) + 2.5\sin(.4T_j)y_2(j-1)]$$
$$\quad + [1 - e^{-.016T_j}\sin(.4T_j + 1.531)]u_{j-1}$$

$$y_2(j) = -e^{-.016T_j}[.4\sin(.4T_j)y_1(j-1) \qquad (5\text{-}13)$$
$$\quad + \sin(.4T_j - 1.531)y_2(j-1)]$$
$$\quad + .4e^{-.016T_j}\sin(.4T_j)u_{j-1}, \qquad j = 1, \ldots, N$$

The performance index chosen in this case is as follows:

$$\text{Minimize} \quad J = \sum_{j=1}^{N-1}[y_1^2(j) + y_2^2(j)] \qquad (5\text{-}14)$$

Additional constraints on the sampling time are imposed as

$$\sum_{i=1}^{N} T_i \leqslant t_{max} = .0305 \text{ sec} \qquad (5\text{-}15)$$

and

$$T_i \geqslant T_{min} = 10^{-3} \text{ sec}, \qquad i = 1, \ldots, N \qquad (5\text{-}16)$$

It should be noted that the constraints given in Eqs. (5-15) and (5-16) operate on the sampling times exclusively. A similar problem has been attacked by Volz [11]. In that paper, necessary conditions for optimality were derived.

The inclusion of constraints on sampling times does not present any computational difficulty with the NLP approach. These constraints are handled just as any others without introducing extra complications into the problem.

The following problem is to be solved numerically: Find the optimal switching times, t_1, t_2, t_3, and t_4, with u attaining the successive values of 1, 2, 1, 2 at the corresponding instants so that the system is brought from $y_1(t_0) = 0$ to $y_1(t_4)$ within 1% of 2, without overshoot; i.e., $y_1(t_4) = 1.98$, minimizing Eq. (5-14), subject to constraints in Eqs. (5-13), (5-15), and (5-16). Therefore,

$$y_1(j) \leqslant 2, \qquad j = 1, 2, 3 \qquad (5\text{-}17)$$

The value of $y_2(t_f)$ is set to be .001. Using the same type of notation as in Chapter 4, we assign the control-problem variables to be specific MP-problem variables.

Control-problem variables	MP variables
T_1	$x(1)$
T_2	$x(2)$
T_3	$x(3)$
T_4	$x(4)$
$y_1(1)$	$x(5)$
$y_2(1)$	$x(6)$
$y_1(2)$	$x(7)$
$y_2(2)$	$x(8)$
$y_1(3)$	$x(9)$
$y_2(3)$	$x(10)$

The MP variables are substituted into the performance index in Eq. (5-14) and into the constraints in Eqs. (5-13) and (5-15) through (5-17). The following MP problem results:

$$\min \left\{ \sum_{j=5}^{10} x^2(j) \,\middle|\, x(i) \geqslant 10^{-3},\, i = 1, 2, 3, 4;\, \sum_{i=1}^{4} x(i) \leqslant .0305; \right.$$

$x(5) \leqslant 2;\, x(7) \leqslant 2;\, x(9) \leqslant 2$

$x(5) = 1 - \exp[-.016x(1)] \sin[.4x(1) + 1.531]$

$x(6) = .4 \exp[-.016x(1)] \sin[.4x(1)]$

$x(7) = \exp[-.016x(2)]\{\sin[.4x(2) + 1.531]x(5) + 2.5 \sin[.4x(2)]x(6)\}$
$\qquad + 2\{1 - \exp[-.016x(2)] \sin[.4x(2) + 1.531]\}$

$x(8) = -\exp[-.016x(2)]\{.4 \sin[.4x(2)]x(5) + \sin[.4x(2) - 1.531]x(6)\}$
$\qquad + .8 \exp[-.016x(2)] \sin[.4x(2)]$

$x(9) = \exp[-.016x(3)]\{\sin[.4x(3) + 1.531]x(7) + 2.5 \sin[.4x(3)]x(8)\}$
$\qquad + \{1 - \exp[-.016x(3)] \sin[.4x(3) + 1.531]\}$

$x(10) = -\exp[-.016x(3)]\{.4 \sin[.4x(3)]x(7) + \sin[.4x(3) - 1.531]x(8)\}$
$\qquad + .4 \exp[-.016x(3)] \sin[.4x(3)]$

$1.98 = \exp[-.016x(4)]\{\sin[.4x(4) + 1.531]x(9) + 2.5 \sin[.4x(4)]x(10)\}$
$\qquad + 2\{1 - \exp[-.016x(4)] \sin[.4x(4) + 1.531]\}$

$.001 = \exp[-.016x(4)]\{.4 \sin[.4x(4)]x(9) + \sin[.4x(4) - 1.531]x(10)\}$
$\qquad \left. + .8 \exp[-.016x(4)] \sin[.4x(4)] \right\}$

Now we have a typical MP problem with ten variables, eight inequality constraints, and eight equality constraints. The problem is in a form suitable for direct implementation of an MP code. The performance index (or the objective function) is quadratic, and the inequality constraints are linear; however, the equality constraints are nonlinear. Therefore we have an NLP problem. Further handling of the format of the equations would depend on the particular MP code used. In this case, the SUMT code, discussed in Chapter 3, was used. A more detailed example of utilizing the SUMT code is given in Sec. 6.4. The results obtained after applying the SUMT code to the solution of the problem were as follows:

i	$T_i m$ (sec)	$t_i m$ (sec)	y_1	y_2	u
1	11.99	11.99	.561	.550	1.0
2	2.42	14.41	.736	.550	2.0
3	1.00	15.41	.471	.678	1.0
4	5.95	21.36	1.980	.001	2.0

The optimal performance index is obtained as

$$J_{\text{opt}} = 2.15$$

The problem was run on an IBM 7094 Computer, entirely coded in FORTRAN IV. The computing time was approximately 1.5 min. This is a typical computing time for the SUMT code for a problem of this size and complexity. It should be pointed out that if the maximum principle were used, a most difficult TPBVP problem would be generated. The computational realization of constraints on the state variables presents a formidable problem. So far, few suitable programs have been written. However, in the MP approach, constraints such as these present no particular problem.

5.4 FUEL-OPTIMAL PROBLEM

A fuel-optimal-control problem is considered in this section [1, 2, 4]. In particular, we consider a class of problems which lend themselves to an LP formulation. Taking into account the formulation in Eqs. (5-3) and (5-10), we can formulate the following problem:

$$\min \left\{ \sum_{i=0}^{N-1} \sum_{j=1}^{M} |u_j(i)| \,\middle|\, A_u \mathbf{u}(i) \leqslant \mathbf{b}_u; i = 0, 1, \ldots, N-1; \right. \\ \left. A_y \mathbf{y}(i) \leqslant \mathbf{b}_y; i = 1, \ldots, N-1; S\mathbf{y}(N) \leqslant \mathbf{b}_s \right\} \quad (5\text{-}18)$$

where A_u, A_y, and S are $m \times m$, $n \times n$, and $n \times n$ constant matrices, respectively, and $\mathbf{b}_u$, $\mathbf{b}_y$, and $\mathbf{b}_s$ are constant m, n, and n vectors, respectively.

$$T = t_i - t_{i-1} \quad \text{for all} \quad i = 1, \ldots, N$$

The expression for $\mathbf{y}(N)$ in Eq. (5-18) can be formulated as follows. Using Eq. (5-9) and assuming a constant $T_i = T$ for all i, we can write

$$\mathbf{y}(i+1) = e^{AT}\mathbf{y}(i) + \int_{t_i}^{t_{i+1}} e^{A(t_{i+1}-\tau)} B\, d\tau\, \mathbf{u}(i) = E\mathbf{y}(i) + F\mathbf{u}(i)$$

where $E = e^{AT}$ is an $n \times n$ constant matrix and $F = \int_{t_i}^{t_{i+1}} e^{A(t_{i+1}-\tau)} B\, d\tau$ is a constant $n \times m$ matrix. The final state is [4]

$$\mathbf{y}(N) = E^N \mathbf{y}(0) + \sum_{i=1}^{N} E^{i-1} F \mathbf{u}(N-i) \qquad (5\text{-}19)$$

This expression is then substituted into the problem formulated in Eq. (5-18). The state and control constraints in Eq. (5-18) are linear and more general then those of Eq. (5-10).

The unknown variables of the problem formulated in Eq. (5-18) are

$u_j(i), \quad j = 1, \ldots, m; \quad i = 0, 1, \ldots, N-1$ a total of mN variables

The number of linear constraints is

Control inequality constraints	mN
State inequality constraints	$n(N-1)$
Terminal-state inequality constraints	n

a total of $N(n + m)$ constraints.

In many applications, the fuel consumption is directly proportional to the absolute value of the control variable, and this is the reason for the formulation of the performance index in Eq. (5-18). Of course, this performance index is not linear in $u_j(i)$, and therefore LP may not be readily applied to the numerical solution of this class of problems. There are several methods of expressing the performance index of a fuel-optimal problem in a linear form. Two of the methods will be mentioned here.

The Bounding Variables Method

This method was proposed by Fath and Higgins [2]. A new set of auxiliary variables $\mathbf{a}(i)$ ($i = 0, \ldots, N-1$) are introduced so that the following inequality constraints are satisfied:

$$|u_j(i)| \leqslant a_j(i), \quad \begin{matrix} j = 1, \ldots, m \\ i = 0, 1, \ldots, N-1 \end{matrix} \qquad (5\text{-}20)$$

or

$$\left.\begin{matrix} u_j(i) \leqslant a_j(i) \\ -u_j(i) \leqslant a_j(i) \end{matrix}\right\} \quad \begin{matrix} j = 1, \ldots, m \\ i = 0, 1, \ldots, N-1 \end{matrix} \qquad (5\text{-}21)$$

The performance index becomes

$$\min J = \sum_{i=0}^{N-1} \sum_{j=1}^{m} a_j(i) \qquad (5\text{-}22)$$

which is a linear expression in terms of the $a_j(i)$ variables. Using this method would mean adding mN variables and $2mN$ constraints to the problem.

Fath [4] developed another efficient variation of this bounding technique for a numerical solution of this particular class of problems which considerably reduces the computational effort required. According to that formulation of the problem, the total number of constraints would be $2mN$, half of which are of the bounding type:

$$a_j(i) \leqslant a_{max}$$

The Difference Variables Method

This method was utilized by Waespy [1]. The control variables $u_j(i)$ ($j = 1, \ldots, m$; $i = 0, 1, \ldots, N-1$) are decomposed into a pair of nonnegative variables as follows:

$$u_j(i) = u'_j(i) - u''_j(i) \tag{5-23}$$
$$u'_j(i) \geqslant 0, \quad j = 1, \ldots, m$$
$$u''_j(i) \geqslant 0, \quad i = 0, 1, \ldots, N-1$$

The performance index is now

$$\min J = \sum_{i=0}^{N-1} \sum_{j=1}^{m} [u'_j(i) - u''_j(i)] \tag{5-24}$$

Equation (5-23) is substituted into all the constraints of Eq. (5-18). The total number of constraints will not change. However, the total number of variables will double, as in the previous method. In summary, the overall constraints matrix will have the same number of columns as previously. However, the number of rows will be less by $2mN$.

5.5 FUEL-OPTIMAL RENDEZVOUS PROBLEM

As an illustrative example of the MP formulation of a fuel-optimal problem, the optimal rendezvous problem in space will be discussed in this section. The results presented in this section were contributed by Waespy [1].

Near earth, space operations may be divided into several mission phases as follows: launch, orbit injection, orbit transfer, midcourse, and terminal maneuvering. For present purposes, the orbit-transfer phase is considered to be one which transfers the spacecraft between earth orbits differing widely in associated energy—altitude differences of hundreds of nautical miles, for example. After execution of such a transfer maneuver, the spacecraft's velocity and position at a specified time may still differ from a desired position and velocity. In the terminal-maneuver phase, we shall assume this difference to be anywhere from 1 to 10 nautical miles and 100–300 ft/sec, depending on a great many mission factors.

Midcourse and terminal maneuvers are consequently required to

1. Bring the spacecraft to a prespecified position, velocity, and time state so that its subsequent free-fall trajectory follows a desired ephemeris.
2. Match the position and velocity of another orbiting spacecraft in order to achieve a rendezvous or station-keeping condition.

Terminal maneuvers are also needed to maintain a desired ephemeris in the face of conservative and nonconservative perturbation forces.

The following assumptions are postulated:

1. Thrust is assumed to be obtained by fixed thrusters mounted orthogonally.
2. Discrete-time-linearized equations of motion (of the Wiltshire-Clohessy type [12]) are employed.
3. A target-centered coordinate system is postulated. In this system the linearized equations for the applicable relative distances are sufficiently accurate.
4. Fuel usage for discrete-time control functions is minimized.
5. Initial and target orbits may be circular or elliptical and may involve plane changes within allowable thrust acceleration, maneuver time, and linearization ranges.
6. The state determination (navigation) system is assumed to be sufficiently precise to make plausible a fuel-minimization problem which is deterministic.

Assuming a circular orbit, we have the following system equations:

$$\ddot{x} - 2w\dot{y} = F_x$$
$$\ddot{y} + 2w\dot{x} - 3w^2 y = F_y \qquad (5\text{-}25)$$
$$\ddot{z} + w^2 z = F_z$$

where $x, y, z =$ Cartesian coordinates of the controlled satellite in a target-centered coordinate system
$w^2 = \mu/R_T^3$
$\mu =$ Earth gravitational constant
$R_T =$ Radius of earth-centered circular rotation of the target satellite
$F_x, F_y, F_z =$ Cartesian components of relative thrust acceleration between the controlled and target satellites

These are the rendezvous equations reported by Clohessy and Wiltshire [12].

Consider the first two equations of Eq. (5-25). These can be used to represent in-plane maneuvers independent of the third out-of-plane equation in z. Set $x = y_1$, $\dot{y}_1 = y_2$, $y = y_3$, $\dot{y}_3 = y_4$, $F_x = u_1$, and $F_y = u_2$ to obtain

$$\begin{bmatrix} \dot{y}_1 \\ \dot{y}_2 \\ \dot{y}_3 \\ \dot{y}_4 \end{bmatrix} = \begin{bmatrix} 0 & 1 & 0 & 0 \\ 0 & 0 & 0 & 2w \\ 0 & 0 & 0 & 1 \\ 0 & -2w & +3w^2 & 0 \end{bmatrix} \begin{bmatrix} y_1 \\ y_2 \\ y_3 \\ y_4 \end{bmatrix} + \begin{bmatrix} 0 & 0 \\ 1 & 0 \\ 0 & 0 \\ 0 & 1 \end{bmatrix} \begin{bmatrix} u_1 \\ u_2 \end{bmatrix} \qquad (5\text{-}26)$$

Sec. 5.5   Fuel-Optimal Rendezvous Problem   81

In like manner, for the equation in z, set $z = y_5$, $\dot{x}_5 = y_6$, and $u_3 = F_z$ to obtain

$$\begin{bmatrix} \dot{y}_5 \\ \dot{y}_6 \end{bmatrix} = \begin{bmatrix} 0 & 1 \\ w^2 & 0 \end{bmatrix} \begin{bmatrix} y_5 \\ y_6 \end{bmatrix} + \begin{bmatrix} 0 \\ 1 \end{bmatrix} u_3 \qquad (5\text{-}27)$$

The in-plane state equations are in the form

$$\dot{\mathbf{y}} = F\mathbf{y} + B\mathbf{u}$$

where F and B are constant matrices. The state transition matrix is obtained as follows:

$$\Phi(t) = \mathscr{L}^{-1}[Is - F]^{-1}$$

where $\mathscr{L}^{-1}$ denotes the inverse Laplace transform operation and I is the identity matrix [9].

Hence

$$\Phi(t) = \begin{bmatrix} 1 & \frac{4}{w}\sin wt - (3t) & 6wt - 6\sin wt & \frac{2}{w} - \frac{2}{w}\cos wt \\ 0 & 4\cos wt - (3) & 6w - 6w\cos wt & 2\sin wt \\ 0 & \frac{2}{w}\cos wt - \left(\frac{2}{w}\right) & 4 - 3\cos wt & \frac{1}{w}\sin wt \\ 0 & -2\sin wt & 3w\sin wt & \cos wt \end{bmatrix} \qquad (5\text{-}28)$$

The applicable difference equations can be formulated from Eqs. (5-26) and (5-28). For the application at hand,

$$\mathbf{y}(k+1) = A\mathbf{y}(k) + B\mathbf{u}(k), \qquad k = 0, 1, \ldots, N-1 \qquad (5\text{-}29)$$

where

$$A = \Phi(T) = \begin{bmatrix} 1 & \frac{4}{w}\sin wT - (3T) & 6wT - 6\sin wT & \frac{2}{w} - \frac{2}{w}\cos wT \\ 0 & 4\cos wT - 3 & 6w - 6w\cos wT & 2\sin wT \\ 0 & \frac{2}{w}\cos wT - \frac{2}{w} & 4 - 3wT & \frac{1}{w}\sin wT \\ 0 & -2\sin wT & 3w\sin wT & \cos wT \end{bmatrix} \qquad (5\text{-}30)$$

$$B = \begin{bmatrix} -\frac{4}{w^2}\cos wT - \left(\frac{3T^2}{2}\right) + \frac{4}{w^2} & \frac{2T}{w} - \frac{2}{w^2}\sin wT \\ \frac{4}{w}\sin wT - (3T) & -\frac{2}{w}\cos wT + \left(\frac{2}{w}\right) \\ \frac{2}{w^2}\sin wT - \left(\frac{2T}{w}\right) & -\frac{1}{w^2}\cos wT + \left(\frac{1}{w^2}\right) \\ \frac{2}{w}\cos wT - \left(\frac{2}{w}\right) & \frac{1}{w}\sin wT \end{bmatrix} \qquad (5\text{-}31)$$

$$\mathbf{u}(k) = [u_1(k), u_2(k)]$$

82 Linear Continuous-Time Systems   Ch. 5

The computer results to follow are based on a target state corresponding to a 300-NM circular orbit. The sampling interval is chosen to be $T = 20$ sec. With these parameters, $wT = .022$ rad and

$$A = \begin{bmatrix} 1.000 & 19.994 & .000 & .438 \\ 0 & .999 & .000 & .044 \\ 0 & -.438 & 1.000 & 19.998 \\ 0 & -.044 & .000 & 1.000 \end{bmatrix} \quad (5\text{-}32)$$

$$B = \begin{bmatrix} 199.969 & 2.923 \\ 19.994 & .438 \\ -2.923 & 199.992 \\ -.438 & 19.998 \end{bmatrix} \quad (5\text{-}33)$$

The initial conditions $\mathbf{y}(0) = [y_1(0), y_2(0), y_3(0), y_4(0)]^T$ and the final conditions $\mathbf{y}(N) = [0, 0, 0, 0]^T$ are specified.

From Eq. (5-19) we can derive (interchanging $A \equiv E$ and $B \equiv F$)

$$-\mathbf{y}(0) = \sum_{j=0}^{N-1} A^{-1-j} B \mathbf{u}(j) \quad (5\text{-}34)$$

$$R(j) = A^{-1-j} B \quad (5\text{-}35)$$

so that

$$R(j+1) = A^{-1} R(j) \quad (5\text{-}36)$$

The LP problem then becomes

1. Find $\mathbf{u}(k) = [u_1(k), u_2(k)]^T$, $k = 0, 1, \ldots, N-1$
2. Subject to

$$-\mathbf{y}(0) = \sum_{j=0}^{N-1} R(j) \mathbf{u}(j); \; |u_i(j)| \leq M; \; i = 1, 2 \quad (5\text{-}37)$$

3. To minimize

$$\text{Fuel consumption} = \sum_{k=0}^{N-1} \{|u_1(k)| + |u_2(k)|\} \quad (5\text{-}38)$$

Although the constraints in Eq. (5-37) are linear, the problem is still not suitable for an LP computation since the objective function in Eq. (5-38) is of the absolute value type. To circumvent this difficulty, we adopt the difference variables method as stated in Eqs. (5-23) and (5-24). Namely, we introduce the following new variables:

$$\begin{matrix} u_i'(k) \geq 0 \\ u_i''(k) \geq 0 \end{matrix} \quad \begin{matrix} k = 0, 1, \ldots, N-1 \\ i = 1, 2 \end{matrix}$$

so that

$$\begin{matrix} u_1(k) = u_1'(k) - u_1''(k) \\ u_2(k) = u_2'(k) - u_2''(k) \end{matrix} \quad k = 0, 1, \ldots, N-1 \quad (5\text{-}39)$$

Substituting Eq. (5-39) into Eq. (5-38), we obtain

$$\sum_{k=0}^{N-1} [u'_1(k) - u''_1(k) + u'_2(k) - u''_2(k)] \qquad (5\text{-}40)$$

The constraints now become

$$-\mathbf{y}(0) = \sum_{j=0}^{N-1} R(j)[\mathbf{u}'(j) - \mathbf{u}''(j)] \qquad (5\text{-}41)$$

$$u'_i(k) - u''_i(k) \leqslant M, \qquad i = 1, 2 \qquad (5\text{-}42)$$

where $k = 0, 1, \ldots, N - 1$.

Now we introduce the LP problem variables:

Control variables	LP variables
$u'_1(0)$	$x(1)$
$u'_2(0)$	$x(2)$
$u'_1(1)$	$x(3)$
$u'_2(1)$	$x(4)$
.	.
.	.
.	.
$u'_1(N-1)$	$x(2N-1)$
$u'_2(N-1)$	$x(2N)$
$u''_1(0)$	$x(2N+1)$
$u''_2(0)$	$x(2N+2)$
$u''_1(1)$	$x(2N+3)$
$u''_2(1)$	$x(2N+4)$
.	.
.	.
.	.
$u''_1(N-1)$	$x(4N-1)$
$u''_2(N-1)$	$x(4N)$

a total of $4N$ variables. The LP problem can now be formulated in its final form:

$$\min \left\{ \sum_{i=1}^{2N} [x(i) - x(2N+i)] \,\middle|\, x(i) \geqslant 0, i = 1, \ldots, 4N; \right.$$

$$x(i) - x(2N+i) \leqslant M, i = 1, \ldots, 2N; \qquad (5\text{-}43)$$

$$\left. -\mathbf{y}(0) = \sum_{j=1} R(j-1) \left[\begin{bmatrix} x(2j-1) \\ x(2j) \end{bmatrix} - \begin{bmatrix} x(2N+2j-1) \\ x(2N+2j) \end{bmatrix} \right] \right\}$$

The problem formulated in Eq. (5-43) is amenable to direct application of any LP code. In this case the LP/90 system [13] was used. The results of the computation are reported as follows:

Figure 5-2 depicts a typical trajectory with relatively high initial relative velocity. The optimal acceleration-time history was computed for 400-sec

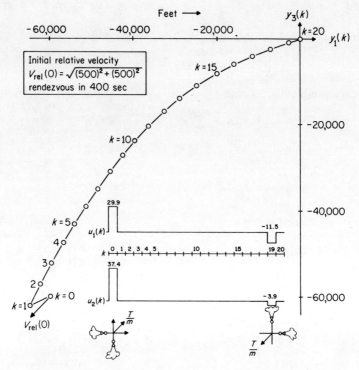

FIG. 5-2. Typical trajectory with high relative velocity [1].

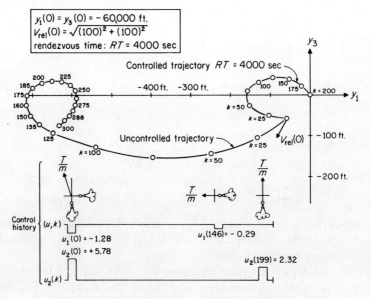

FIG. 5-3. Controlled and uncontrolled trajectories I [1].

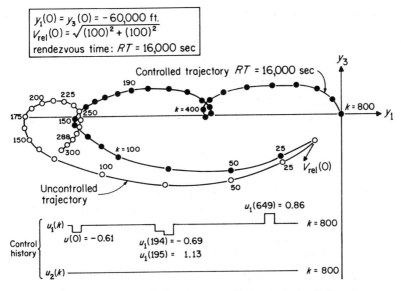

FIG. 5-4. Controlled and uncontrolled trajectories II [1].

rendezvous time with sample time taken as 20 sec. If the sample time were reduced, the pulses shown would be narrower but of higher amplitude. In the limit, the theoretical optimum impulse control history would result.

More realistic initial relative velocities are used for the sample trajectories of Fig. 5-3 and 5-4.

5.6 THE TIME-OPTIMAL PROBLEM

The system considered is described by the following set of state equations [3]:

$$\dot{\mathbf{y}} = A\mathbf{y} + B\mathbf{u} \tag{5-44}$$

where $\mathbf{y}$ = n-dimensional state vector
$\mathbf{u}$ = r-dimensional control vector
A = $n \times n$ constant matrix
B = $n \times r$ constant matrix

The following inequality constraints are imposed on the system at all times:

$$C_u \mathbf{u} \leqslant \mathbf{v}_u(t) \tag{5-45}$$

$$C_y \mathbf{y} \leqslant \mathbf{v}_y(t) \tag{5-46}$$

where $\mathbf{v}_u$ = q-dimensional vector
$\mathbf{v}_y$ = p-dimensional vector
C_u = $q \times r$ constant matrix
C_y = $p \times n$ constant matrix

The target set at the final time $t = t_f$ is defined by the following inequalities:

$$G\mathbf{y}(t_f) \leqslant \mathbf{v}_g \qquad (5\text{-}47)$$

$$C_y \mathbf{y}(t_f) \leqslant \mathbf{v}_y \qquad (5\text{-}48)$$

where $\mathbf{v}_g = g$-dimensional vector
$G = g \times n$ constant matrix

The time-optimal-control problem for the system under consideration is stated as follows:

Given an initial state $\mathbf{y}(0)$, determine the control vector function $\mathbf{u}(t)$ such that

1. $\mathbf{u}(t)$ satisfies the constraints in Eq. (5-45).
2. The system of Eq. (5-44) is taken from $\mathbf{y}(0)$ to some point within the target set given by Eqs. (5-47) and (5-48).
3. The resulting trajectory satisfies the constraint of Eq. (5-46).
4. The final time t_f is minimized.

An algorithm for an approximate solution of this problem, utilizing linear programming, has been proposed by Fath [3]. This algorithm is based on the notion of the *set of attainability*, which is defined next.

Definition 5-1. A *set of attainability*, A_f, is the set of all initial states that can be reached by admissible trajectories backing out from the given target set on the time interval $0 \leqslant t \leqslant t_f$. Or, in other words, A_f is the set of all initial states for which there exist admissible control functions $\mathbf{u}(t)$ on the interval $0 \leqslant t \leqslant t_f$ that produce admissible trajectories terminating within the specified target set. *Admissible* control functions and trajectories are those satisfying all of their respective constraints.

The minimum time t_f in the time-optimal-control problem stated above corresponds to the time at which $\mathbf{y}(0)$ is first in the set of attainability A_f. By investigating some of the properties of the set A_f for the class of systems considered in the section, insight into the time-optimal-control problem can be obtained.

As the system is linear and the set of admissible control functions bounded (which results in bounded derivatives of the state variables), the set of attainability is continuous with respect to t_f. Thus the minimum time for solution occurs when $\mathbf{y}(0)$ is first on the boundary of A_f. This also means that if at any time t_{f1}, $\mathbf{y}(0)$ is exterior to the set A_{f1}, and if at some other time $t_{f2} > t_{f1}$, $\mathbf{y}(0)$ is interior to A_{f2}, then at some intermediate time, t_f, $\mathbf{y}(0)$ must be on the boundary of A_f.

A second important property of the set of attainability is that it is at all times convex. This follows from the convexity of the constraint regions and the linearity of the system. A complete proof is given in reference 3.

A third property possessed by the set of attainability under certain conditions is the *inclusion* property which is described below.

Definition 5-2. A set of attainability is said to satisfy an *inclusion* relation (and thus possesses the inclusion property) on an interval $0 \leqslant t \leqslant t_{f2}$ if whenever any initial state $\mathbf{y}(0)$ is an element of the set of attainability at a time t_{f1}, it is an element of the set of attainability for all t_f such that

$$t_{f1} \leqslant t_f \leqslant t_{f2}$$

This property indicates a nondecreasing set of attainability. Although this property is possessed by sets of attainability for many systems, it is not a general property of all sets of attainability. All systems with the target set as the origin in state space do possess this property, but for general target sets, other conditions must be imposed to guarantee satisfaction of the inclusion relation.

Sufficient conditions that guarantee that the set of attainability satisfies the inclusion relation are given by the following theorem, the proof of which is given in reference 3.

Theorem 5-1. If on the interval $0 \leqslant t \leqslant t_{f2}$ for any two times t_{f0} and t_{f1} such that

$$0 \leqslant t_{f0} \leqslant t_{f1} \leqslant t_{f2} \tag{5-49}$$

and for any state $\mathbf{y}(t_{f0})$ in the target set, there exists an admissible control function of the interval $t_{f0} \leqslant t \leqslant t_{f1}$ such that $\mathbf{y}(t)$, given by

$$\mathbf{y}(t) = e^{A(t-t_{f0})}\mathbf{y}(t_{f0}) + \int_{t_{f0}}^{t} e^{A(t-\sigma)} \mathbf{B}\mathbf{u}(\sigma)\, d\sigma \tag{5-50}$$

(where σ is an integration variable representing time),

1. is an admissible trajectory on $t_{f0} \leqslant t \leqslant t_{f1}$
2. $\mathbf{y}(t_{f1})$ is an element of the target set

then the set of attainability satisfies an inclusion relation on the interval $0 \leqslant t \leqslant t_{f2}$.

It should be remarked that the conditions are sufficient but by no means necessary. Many sets of attainability have this property even though the target set restrictions given above are not met. Further, if the conditions of the theorem are met by more than one control, then the system is said to satisfy an inclusion relation of the second kind.

Under the assumption that the set of attainability satisfies an inclusion relation of the second kind, it is easy to see that if an initial state on the boundary of the set of attainability is reached by a unique trajectory, then the trajectory must be time optimal. If the state were in the set of attainability at any time sooner, by assumption the trajectory would not be unique. In reality, this situation is usually the case. The time-optimal trajectory is usually unique (although it is by no means necessary), and all points on the boundary of the set A_f that are not the result of time-optimal trajectories are reached by multiple trajectories. This fact is of importance in the proposed approximation scheme.

The principal idea behind this approximation procedure is reformulation of the problem such that on a fixed-time interval, the trajectory backing out from the target set to the intersection of a specified line and the boundary of the set of attainability can be obtained. By requiring the specified line to pass through the origin and the desired initial state y(0) and by adjusting the length of the time interval, a trajectory (when one exists) resulting in y(0) being on the boundary of the set of attainability can be obtained. Conditions then must be checked to determine if the resulting trajectory is actually time optimal.

The first step in the approximation procedure is to divide the time interval $[0, t_f]$ into N subintervals of length T, where

$$T = \frac{t_f}{N} \tag{5-51}$$

Next, it is assumed that the control vector $\mathbf{u}(t)$ has constant components on each of the subintervals. This assumption essentially means that a stepwise approximation to the actual optimal control is being sought. With this assumption, let the sequence of constant control vectors $\mathbf{u}(k)$ be defined as

$$\mathbf{u}(k) = \mathbf{u}(t), \quad \begin{array}{c} (k-1)T \leqslant t < kT \\ k = 1, 2, \ldots, N \end{array} \tag{5-52}$$

The one other assumption made in this approximation procedure is that the state constraints and the control constraints are enforced only at discrete instants of time—once within each subinterval. Thus the state constraints are represented as

$$C_y \mathbf{y}(kT) \leqslant \min \mathbf{v}_y(t) \equiv \mathbf{v}_y(k), \quad \begin{array}{c} kT \leqslant t \leqslant (k+1)T \\ k = 0, 1, \ldots, N-1 \end{array} \tag{5-53}$$

This formulation enforces the state constraints at the most restrictive point in the subinterval.

The representation of the control constraints is divided into two parts. The first part includes the constraints in the form of simple bounds on the control vector during a subinterval and is given by

$$\mathbf{u}(k) \leqslant \mathbf{V}_u(k), \quad k = 1, 2, \ldots, N \tag{5-54}$$

$$\mathbf{u}(k) \geqslant \mathbf{V}_L(k), \quad k = 1, 2, \ldots, N \tag{5-55}$$

where $\mathbf{V}_u(k)$ and $\mathbf{V}_L(k)$ are the upper and lower limits, respectively, for $\mathbf{u}(k)$ on the kth subinterval. As the admissible region for the controls is assumed closed and bounded, constraints in the form of Eqs. (5-54) and (5-55) either are present or can be imposed in every problem. Having constraints in this form is advantageous in the proposed computing algorithm. The second part of the control constraint formulation includes all the constraints involving linear combinations of the control vector components and is represented by

$$C_u \mathbf{u}(k) \leqslant \mathbf{v}_u(k), \qquad k = 1, 2, \ldots, N \tag{5-56}$$

where C_u is a $(q \times r)$-dimensional constraint matrix and $\mathbf{v}_u(k)$ is the q-dimensional constraint vector for the kth subinterval [see Eq. (5-45)].

The target set description given in Eqs. (5-47) and (5-48) is likewise separated into two sections. The first consists of the simple bounds

$$\mathbf{y}(t_f) \leqslant \bar{\mathbf{X}}_u \tag{5-57}$$

$$\mathbf{y}(t_f) \geqslant \bar{\mathbf{X}}_l \tag{5-58}$$

where $\bar{\mathbf{X}}_u$ and $\bar{\mathbf{X}}_l$ are n-dimensional vectors representing the upper and lower limits, respectively, for $\mathbf{y}(t_f)$. The second part of the target set constraint consists of constraints on the linear combination of the components of $\mathbf{y}(t_f)$ and is represented by [see also Eq. (5-47)]

$$G\mathbf{y}(t_f) \leqslant \mathbf{v}_g \tag{5-59}$$

where G is a $(g \times n)$-dimensional constraint matrix and $\mathbf{v}_g$ is a g-dimensional constraint vector.

The final constraint set to be considered is that which requires the initial state $\mathbf{y}(0)$, resulting from a trajectory backing out from the target set to lie on a specified line in state space. Let $\mathbf{P}$ be an n-dimensional unit vector pointing toward $\mathbf{y}(0)$. A line through the origin in the direction of $\mathbf{P}$ can be specified as the intersection of $n - 1$ hyperplanes. This constraint can thus be presented by

$$E\mathbf{y}(0) = \mathbf{0} \tag{5-60}$$

where E is an $[(n - 1) \times n]$-dimensional constraint matrix. The components of E are not unique, as the only condition that must hold is that the vector $\mathbf{P}$ must lie in each hyperplane described by a row of E. That is,

$$E\mathbf{P} = \mathbf{0} \tag{5-61}$$

Thus any linearly independent set of rows for E such that Eq. (5-61) holds is a valid choice.

The object now is to place all the constraints on the sequence of control vectors $\mathbf{u}(k)$ and the final state $\mathbf{y}(t_f)$. This is possible as the state of the system at any instant jT can be expressed in terms of the sequence of control vectors and the final state as

$$\mathbf{y}(jT) = e^{-A(N-j)T}\mathbf{y}(t_f) + \sum_{k=j+1}^{N}\left[-e^{-A(k-j)T}\int_{-T}^{0} e^{-As}B\,ds\right]\mathbf{u}(k) \tag{5-62}$$

which, after defining the multiplier of $\mathbf{y}(t_f)$ as $E_A(N - j)$ and the multiplier of $\mathbf{u}(k)$ as $D(k - j)$, can be written as

$$\mathbf{y}(jT) = E_A(N-j)\mathbf{y}(t_f) + \sum_{k=j+1}^{N} D(k-j)\mathbf{u}(k) \tag{5-63}$$

Using Eq. (5-63), the state constraints of Eq. (5-53) become

$$C_y E_A(N-j)\mathbf{y}(t_f) + \sum_{k=j+1}^{N} C_y D(k-j)\mathbf{u}(k) \leqslant \mathbf{v}_y(j), \\ j = 0, 1, 2, \ldots, N-1 \tag{5-64}$$

and the line constraint of Eq. (5-60) becomes

$$E E_A(N)\mathbf{y}(t_f) + \sum_{k=1}^{N} E D(k)\mathbf{u}(k) \leqslant 0 \tag{5-65}$$

$$-E E_A(N)\mathbf{y}(t_f) - \sum_{k=1}^{N} E D(k)\mathbf{u}(k) \leqslant 0 \tag{5-66}$$

where the two conditions are used to convert the equality constraints to inequality constraints.

The entire set of constraints [Eqs. (5-54) through (5-59) and (5-64) through (5-66)] can now be combined into one matrix inequality involving an $(N \cdot r + n)$-dimensional vector variable $\mathbf{u}_c$, where

$$\mathbf{u}_c = \begin{bmatrix} \mathbf{u}(N) \\ \mathbf{u}(N-1) \\ \cdot \\ \cdot \\ \cdot \\ \mathbf{u}(1) \\ \mathbf{x}(t_f) \end{bmatrix} \tag{5-67}$$

Let this constraint inequality be represented as

$$A_t \mathbf{u}_c \leqslant \mathbf{u}_A \tag{5-68}$$

Any values for the vector $\mathbf{u}_c$ subject to the constraint of Eq. (5-68) result in admissible controls and valid final states that give rise to admissible trajectories terminating at initial states within the set of attainability A_f and on the line specified by Eq. (5-60).

To obtain the set of values for $\mathbf{u}_c$ that result in the initial state being on the boundary of the set of attainability, it is necessary and sufficient that a functional z be maximized, where

$$\begin{aligned} z &= \mathbf{P}^T \mathbf{y}(0) \\ &= \mathbf{P}^T E_A(N) \mathbf{y}(t_f) + \sum_{k=1}^{N} \mathbf{P}^T D(k)\mathbf{u}(k) \end{aligned} \tag{5-69}$$

which can be represented as

$$z = \mathbf{C}^T \mathbf{u}_c \tag{5-70}$$

For a given value of z, Eq. (5-69) is the representation of a hyperplane orthogonal to $\mathbf{P}$ in state space. Maximizing z subject to Eq. (5-68) thus finds the control sequence and final state that result in $\mathbf{y}(0)$ being the maximum distance from the origin and still within the set of attainability. As the set of attainability is convex, this state is well defined.

The problem of determining $\mathbf{u}_c$ resulting in a trajectory leading to the intersection of a line and the boundary of the set of attainability is thus a LP problem maximizing Eq. (5-70) subject to Eq. (5-68). Considering the inequality constraints and the unrestricted sign of the components of $\mathbf{u}_c$, it is natural to consider the above LP problem as a dual LP problem (see Chapter 2) and to obtain its solution through the solution of the corresponding primal LP problem, given by the following: Minimize $y = \mathbf{u}_A^T \mathbf{w}$ subject to

$$A_t^T \mathbf{w} = \mathbf{C}, \quad \mathbf{w} \geqslant 0 \qquad (5\text{-}71)$$

where $\mathbf{w}$ is a dual vector variable. Duality theory in linear programming allows the solution $\mathbf{u}_c$ to be obtained once the optimum $\mathbf{w}$ has been found. The problem of Eq. (5-71) can be solved using the simplex method (see Chapter 3).

Using the above procedure and a fixed-time interval $[0, t_f]$, the intersection of line and the set of attainability can be found. The relation of this point to the desired initial state $\mathbf{y}(0)$ can be determined by comparing the scalar value of the objective function z to the desired value z^*, where

$$z^* = \mathbf{P}^T \mathbf{y}(0) \qquad (5\text{-}72)$$

For z greater than z^*, the desired initial state is interior to the set of attainability, so a shorter time interval must be used. For z less than z^*, the time interval must be lengthened. One good way to adjust the final time is by linear interpolation. Let t_{f1} be the end point of a time interval where $z_1 < z^*$ and let t_{f2} be the end point of another time interval where $z_2 > z^*$. Then a choice of a new final time t_f can be

$$t_f = t_{f1} + \left(\frac{z^* - z_1}{z_1 - z_2}\right)(t_{f2} - t_{f1}) \qquad (5\text{-}73)$$

Using the above iteration on the final time, either a final time can be found with a corresponding value of z within an error tolerance of z^*, or a point will be reached where further increases in t_f do not result in increases in z. The latter case (usually indicated by nonuniqueness of the solution) occurs when the desired initial state $\mathbf{y}(0)$ is outside the region where solutions exist, i.e., outside the set of attainability for an infinite time interval.

Thus, where they exist, the time interval, the final state, the control sequence, and the trajectory resulting in $\mathbf{y}(0)$ being on the boundary of the set of attainability can be determined. What remains is to verify that the approximation is time optimal. Experimentally this can be done by checking time intervals shorter than the optimum time interval and verifying that the resulting values of z are less than z^*. Where it can be verified that the set of attainability satisfies an inclusion relation of the second kind, then uniqueness of the results (indicated in the LP solution) guarantees time optimality. As an example, consider the fourth-order system described by

$$\begin{bmatrix} \dot{y}_1 \\ \dot{y}_2 \\ \dot{y}_3 \\ \dot{y}_4 \end{bmatrix} = \begin{bmatrix} 0 & 1 & 0 & 0 \\ -2 & 0 & 1 & 1 \\ 0 & 0 & 0 & 1 \\ 1 & 0 & -1 & 0 \end{bmatrix} \begin{bmatrix} y_1 \\ y_2 \\ y_3 \\ y_4 \end{bmatrix} + \begin{bmatrix} 0 & 0 \\ 1 & 0 \\ 0 & 0 \\ 0 & 1 \end{bmatrix} \begin{bmatrix} u_1 \\ u_2 \end{bmatrix} \quad (5\text{-}74)$$

Let the control vector be subject to the constraints

$$.2e^{-t} - 1.3 \leqslant u_1(t) \leqslant 1.0 - .2e^{-t}$$
$$.2e^{-t} - 1.2 \leqslant u_2(t) \leqslant 1.1 - .2e^{-t} \quad (5\text{-}75)$$
$$|u_1(t)| + |u_2(t)| \leqslant 1.0 - e^{-t}$$

The above constraints are chosen to indicate the generality of the constraint sets considered. Notice that time-varying unsymmetrical bounds are placed on the control vector components as well as a time-varying bound on the sum of their absolute values.

Let the state constraints be given by

$$\left.\begin{array}{l} -.6 \leqslant y_2 \leqslant .7 \\ -.6 \leqslant y_4 \leqslant .7 \end{array}\right\} \quad 0 \leqslant t \leqslant 1$$
$$\left.\begin{array}{l} -.5 \leqslant y_2 \leqslant .5 \\ -.5 \leqslant y_4 \leqslant .5 \end{array}\right\} \quad 1 < t \quad (5\text{-}76)$$

This constraint set is both time varying and discontinuous.

Let the target set be specified by

$$-.02 \leqslant y_1(t_f) \leqslant .02$$
$$y_2(t_f) = 0$$
$$-.05 \leqslant y_3(t_f) \leqslant .05 \quad (5\text{-}77)$$
$$y_4(t_f) = 0$$
$$|2y_1(t_f)| + |y_3(t_f)| \leqslant .06$$

This formulation includes cases where some components are specified exactly and where others are subject to both simple and combined constraints. A more detailed rearrangement of the problem, suitable for a direct implementation of the simplex algorithm, is left as an exercise for the reader.

The time-optimal problem of interest is to find an allowable control which takes the system from

$$\mathbf{y}(0) = \begin{bmatrix} .50 \\ 0 \\ .75 \\ 0 \end{bmatrix} \quad (5\text{-}78)$$

to the target set in a minimum time subject to both control and state constraints.

Using a computer implementation of the procedures presented, basically utilizing the simplex algorithm (Chapter 3), the results given in Figs. 5-5 and 5-6 were obtained when a 20-step approximation was specified. The stepwise

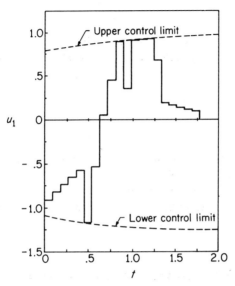

FIG. 5-5. Approximation to the first component of the time-optimal control vector [3].

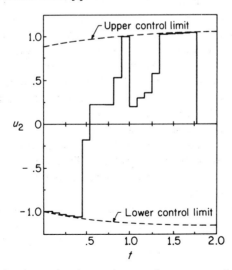

FIG. 5-6. Approximation to the second component of the time-optimal control vector [3].

approximations for u_1 and u_2 are given in Figs. 5-5 and 5-6, respectively. The final state chosen was

$$\mathbf{y}(t_f) = \begin{bmatrix} .005 \\ 0 \\ .05 \\ 0 \end{bmatrix} \quad (5\text{-}79)$$

which is on the boundary of the target set. By allowing an error of .002 in the value of the objective function value, the resulting initial state found was

$$\mathbf{y}(0) = \begin{bmatrix} .50128 \\ 0 \\ .75191 \\ 0 \end{bmatrix} \quad (5\text{-}80)$$

which is "close" to the desired value. The time interval selected was [0, 1.768], which is a minimum in light of the fact that the solution was unique, and the set of attainability satisfies an inclusion relation of the second kind. (It can be shown that it is sufficient to check the conditions of Theorem 5-1 at the extreme points of the convex target set.)

5.7 THE GENERALIZED-PROGRAMMING APPROACH

The LP problem was formulated in Chapters 2 and 3. There exists a variation of the LP problem, called the generalized-programming problem, which has been described by Dantzig [14, chap. 22]. In a GP problem, some of the columns of the constraint coefficients matrix A may be variables of the problem, in which case they are to be solved for. The GP problem is particularly applicable to the formulation of a MP solution of an optimal-control problem for a linear system. This will be discussed in more detail in this section. The idea of applying GP in linear optimal-control problems was originated by Dantzig [15] and discussed in the theses of Waespy [1] and Van Slyke [16]. However, neither man presented a direct numerical solution to a GP formulation of an optimal-control problem. This kind of solution recently has been achieved by Jizmagian [7, 19], and the highlights of his algorithm will be described in this section.

Initially, the GP problem will be defined, and then its particular application to the solution of optimal-control problems will be discussed.

The GP problem can be formulated as follows [7, 14, 19]: Choose a vector $\mathbf{P}$ in a convex set $C \subseteq E^n$ such that we

Maximize z

Subject to $\mathbf{U}_0 z + \mathbf{P}\mu = \mathbf{S}, \quad \mu = 1, \quad \mu \geqslant 0$ \quad (5-81)

Sec. 5.7   The Generalized-Programming Approach   95

where $\mathbf{U}_0$ and $\mathbf{S}$ are specified n-dimensional vectors and μ is a scalar.

Thus, we are looking for some vector $\mathbf{P}^*$ or a convex combination of vectors $\mathbf{P}^{i*}$, all in set C, so that the linear equations are feasible; i.e.,

$$\mathbf{U}_0 z + \mathbf{P}^* = \mathbf{S} \tag{5-82}$$

or

$$\mathbf{U}_0 z + \sum_i \mathbf{P}^{i*} \mu_i = \mathbf{S} \tag{5-83}$$
$$\sum \mu_i = 1, \quad \mu_i \geqslant 0$$

and the resulting value of z is a maximum over the choice of all the elements in set C which satisfy the linear equations. Note that if any set of $\mathbf{P}^i$ is in C, any convex combination of that set is also in C. Hence Eqs. (5-82) and (5-83) are equivalent when

$$\mathbf{P}^* = \sum_i \mathbf{P}^{i*} \mu_i$$
$$\sum \mu_i = 1, \quad \mu_i \geqslant 0$$

The solution procedure is based on the assumption that we have on hand, initially, n particular choices of $\mathbf{P}^i \in C$ so that the following linear program, called a *restricted master program*,

$$\max_\mu z$$

Subject to $\mathbf{U}_0 z + \mathbf{P}^1 \mu_1 + \cdots + \mathbf{P}^n \mu_n = \mathbf{S}$ (5-84)

$$\mu_1 + \cdots + \mu_n = 1, \quad \mu_i \geqslant 0$$

has a unique, feasible, nondegenerate solution with the basis being defined as

$$B^0 = \begin{bmatrix} \mathbf{U}_0 & \mathbf{P}^1 & \cdots & \mathbf{P}^n \\ 0 & 1 & \cdots & 1 \end{bmatrix}$$

and being nonsingular (by definition). Since for each $\mathbf{P}^i \in C$,

$$\mathbf{P}^0 = \sum_i \mathbf{P}^i \mu_i^0$$

where μ_i^0 is a solution to the problem in Eq. (5-84), is in C, and is a feasible solution to Eq. (5-81)—but not necessarily the optimal solution.

To test $\mathbf{P}^0$ [and hence any solution of Eq. (5-81) generated from a basis] for optimality, a row vector $\boldsymbol{\pi} = \boldsymbol{\pi}^0$ is determined to satisfy

$$\boldsymbol{\pi}^0 B^0 = (1, 0, \ldots, 0) \tag{5-85}$$

From $\boldsymbol{\pi}$, we find a vector $\mathbf{P}^{n+1}$, which is not necessarily unique, and a value δ so that

$$\delta = \boldsymbol{\pi}^0 \mathbf{P}^{n+1} = \min_{\mathbf{P} \in C} \boldsymbol{\pi}^0 \bar{\mathbf{P}} \tag{5-86}$$

where

$$\bar{\mathbf{P}} = \begin{bmatrix} \mathbf{P} \\ 1 \end{bmatrix}$$

If $\delta = 0$, the current solution is an optimal one. If $\delta < 0$, Eq. (5-84) is augmented by $\mathbf{P}^{n+1}$, and the new linear program is then solved. The general iteration starts with a solution to the restricted master program:

Maximize z

$$\text{Subject to } \mathbf{U}_0 z + \sum_{i=1}^{n+k} \mathbf{P}^i \mu_i = \mathbf{S} \qquad (5\text{-}87)$$

$$\sum_i \mu_i = 1, \qquad \mu_i \geqslant 0$$

Let B^k be the optimal basis of the linear program in Eq. (5-87), and let $\boldsymbol{\pi}^k$, the dual (optimal) variable to the problem in Eq. (5-87), be defined analogously to Eq. (5-85). Then, δ^k and $\mathbf{p}^{n+k+1}$ are found from the *subproblem*: Find

$$\delta^{k+1} = \boldsymbol{\pi}^k \mathbf{P}^{n+k+1} = \min_{\mathbf{P} \in C} \boldsymbol{\pi}^k \bar{\mathbf{P}} \qquad (5\text{-}88)$$

If $\delta^{k+1} = 0$, the solution of the kth iteration of the master problem is optimal. If $\delta^{k+1} < 0$, then $\mathbf{P}^{n+k+1}$ can be adjoined to Eq. (5-87), and the solution to Eq. (5-81) is improved. The value $-\delta^{k+1}$ is the maximum amount by which the value of the current basis z^k can be improved. Thus, $z - \delta^{k+1}$ constitutes an upperbound to the optimal solution of Eq. (5-81). It is known that these upperbound evaluations can vary considerably from one iteration to the next. Accordingly, the least of these evaluations is saved from all iterations, including the current one.

As will be seen in the forthcoming discussion, the application of GP to the solution of the optimal-control problem generates a subproblem designated as the parametric-programming (PP) problem [7].

A *parametric linear-programming* problem is defined as follows (this is only a particular case; however, it is sufficient for the current implementation):

Find $\mathbf{x}^*(t)$

To minimize $\boldsymbol{\alpha}^T(t)\mathbf{x}$ \qquad (5-89)

Subject to $A\mathbf{x} = \mathbf{b}, \qquad x_i \geqslant 0; i = 1, \ldots, n; t \in [T_1, T_2]$

Similarly, a *parametric quadratic-programming* problem is defined as follows (again, a particular but sufficient case):

Find $\mathbf{x}^*(t)$

To minimize $\boldsymbol{\alpha}^T(t)\mathbf{x} + \mathbf{x}^T D \mathbf{x}$ \qquad (5-90)

Subject to $A\mathbf{x} = \mathbf{b}, \qquad x_i \geqslant 0; i = 1, \ldots, n; \qquad t \in [T_1, T_2]$

In both cases, A is a given $m \times n$ matrix, $\mathbf{b}$ is a given m-dimensional vector, D is an $n \times n$ positive semidefinite matrix (see Chapter 3), $\mathbf{x}$ is an n-dimensional variables vector, and

$$\boldsymbol{\alpha}^T(t) = [\alpha_1(t), \ldots, \alpha_k(t), \ldots, \alpha_n(t)] \qquad (5\text{-}91)$$

is a given vector-valued function, each component of which is a solution to some homogeneous linear differential equation with constant real coefficients that may depend on k. Such $\alpha_k(t)$ are of the form

$$\alpha_k(t) = \sum_{i=1}^{n} P_{ki}(t) e^{s_{ki} t} \qquad (5\text{-}92)$$

where $P_{ki}(t)$ is a polynomial with real coefficients of degree m_{ki} so that

$$\sum_{i=1}^{n} m_{ki} = n \qquad (5\text{-}93)$$

and the s_{ki}'s are constants. If s_{ki} is complex for i odd, s_{ki+1} is its conjugate and $P_{ki}(t) = P_{ki+1}(t)$. It follows, then, that $\alpha_k(t)$ are real-valued functions of t.

The algorithm and basic theorems connected with the solution of this class of problems have been worked out by Jizmagian [7, 20].

The control problems considered will now be formulated as generalized programs, and then the subproblems will be shown to be PP problems. The control problem can be formulated as follows:

$$\text{Minimize } J = \int_0^T \dot{y}_0(t)\, dt = y_0(T), \qquad y_0(0) = 0 \qquad (5\text{-}94)$$

$$\text{Subject to } \dot{\mathbf{y}}(t) = F\mathbf{y}(t) + G\mathbf{u}(t), \qquad \begin{array}{l} \mathbf{y}(0) \in S_0 \text{ (initial set)} \\ \mathbf{y}(T) \in S_T \text{ (target set)} \end{array} \qquad (5\text{-}95)$$

An additional constraint on the control vector is

$$\mathbf{u}(t) \in U = \{\mathbf{u} \mid A\mathbf{u} \geqslant \mathbf{b}\} \qquad (5\text{-}96)$$

where $\mathbf{y} = n$-dimensional vector
$\mathbf{u} = m$-dimensional vector
$F = n \times n$ real matrix
$G = n \times m$ real matrix:

$$\dot{y}_0(t) = \mathbf{f}_0^T \mathbf{y}(t) + \mathbf{g}_0^T \mathbf{u}(t) + h \sum_{i=0}^{m} |u_i(t)| + \mathbf{u}^T(t) Q \mathbf{u}(t) \qquad (5\text{-}97)$$

where $\mathbf{f}_0$ is a fixed real n vector and $\mathbf{g}_0$ is a fixed real m vector, h is a real constant, and Q is an $n \times n$ real matrix.

First, consider strictly linear cost functionals; i.e., $h = 0$ and $Q = 0$. Letting

$$\bar{F} = \begin{bmatrix} 0 & \mathbf{f}_0^T \\ \hline \vdots & \vdots \\ 0 & F \end{bmatrix} \qquad (n+1) \times (n+1) \qquad (5\text{-}98)$$

$$\bar{G} = \begin{bmatrix} \mathbf{g}_0^T \\ \hline G \end{bmatrix} \qquad (n+1) \times m$$

the completely linear system can be expressed by
$$\dot{\bar{\mathbf{y}}}(t) = \bar{F}\bar{\mathbf{y}}(t) + \bar{G}\mathbf{u}(t) \tag{5-99}$$
where
$$\bar{\mathbf{y}} = \begin{bmatrix} y_0 \\ \mathbf{y} \end{bmatrix}$$

When a particular vector function $\mathbf{u}^i(t)$ and an initial condition $\bar{\mathbf{y}}^i(0)$ are given, the solution to (5-99), at $t = T$, is

$$\bar{\mathbf{y}}(T) = e^{T\bar{F}}\bar{\mathbf{y}}^i(0) + \int_0^T e^{(T-t)\bar{F}}\bar{G}\mathbf{u}^i(t)\,dt \tag{5-100}$$

and the solution to (5-95) is

$$\mathbf{y}(T) = e^{TF}\mathbf{y}^i(0) + \int_0^T e^{(T-t)F}G\mathbf{u}^i(t)\,dt \tag{5-101}$$

The final set is redefined as
$$\bar{S}_T = S_T - S_0^F$$
$$S_0^F \equiv \{\mathbf{v} \mid \mathbf{y} \in S_0,\ \mathbf{v} = e^{TF}\mathbf{y}\}$$

It can be shown [7] that if S_0 and S_T are convex, so are S_0^F and $\bar{S}_T$. Without loss of generality, we can assume that
$$\mathbf{y}(0) = 0$$
and $\tag{5-102}$
$$\mathbf{y}_0(0) = 0$$

If we take the vector functionals of the control $\mathbf{P} = \mathbf{P}[u(t)]$ to be defined by

$$\mathbf{P} = \int_0^T e^{(T-t)F}G\mathbf{u}(t)\,dt \tag{5-103}$$

and
$$\bar{\mathbf{P}} = \bar{\mathbf{P}}[u(t)]$$

to be defined by

$$\bar{\mathbf{P}} = \int_0^T e^{(T-t)\bar{F}}\bar{G}\mathbf{u}(t)\,dt \tag{5-104}$$

then let the sets C and $\bar{C}$ be defined by

$$C \equiv \left\{ \mathbf{P} \mid \mathbf{u}(t) \in U,\ \mathbf{P} = \int_0^T e^{(T-t)F}G\mathbf{u}(t)\,dt \right\}$$

and

$$\bar{C} \equiv \left\{ \bar{\mathbf{P}} \mid \mathbf{u}(t) \in U,\ \bar{\mathbf{P}} = \int_0^T e^{(T-t)\bar{F}}\bar{G}\mathbf{u}(t) \right\} dt$$

Let $U_0^T = (1, 0, \ldots, 0)$ and note that the first component of the vectors $\mathbf{S}$, in the set $\bar{S}_T$, is defined to be zero. Also note that the first component of the $\bar{\mathbf{P}}$ vector represents the cost of using the control (and its corresponding trajectory) generating $\bar{\mathbf{P}}$. Thus we are looking for a vector $\bar{\mathbf{P}} \in \bar{C}$, a vector function $\mathbf{u}(t)$ generating $\bar{\mathbf{P}}$, and a vector $\bar{\mathbf{S}} \in \bar{S}_T$ to satisfy

$$\max_{\mathbf{P} \in C} \lambda, \quad \mu, \nu \geqslant 0$$
$$\text{Subject to} \quad \mathbf{U}_0 \lambda + \bar{\mathbf{P}}\mu = \bar{\mathbf{S}}\nu, \quad \begin{aligned} \mu &= 1 \\ \nu &= 1 \end{aligned} \quad (5\text{-}105)$$

where μ and ν are scalars. Maximizing λ is equivalent to minimizing $J[\mathbf{u}(t)]$, the first component of the vector $\bar{\mathbf{P}}$, where $\mathbf{u}(t)$ generates $\bar{\mathbf{P}}$. Since $\bar{\mathbf{P}}$ must be taken from a convex set $\bar{C}$ and $\bar{\mathbf{S}}$ must be taken from a convex set $\bar{S}_T$, the above formulation is a generalized program of the Dantzig-Wolfe type [14].

A solution to the GP problem consists of a vector $\mathbf{P}$ in the reachable set R_T, a control function $\mathbf{u}(t)$ in the admissible control region U generating $\mathbf{P}$, and a vector $\mathbf{S}$ in the constraint set of terminal states $\bar{S}_T$, so that

$$\mathbf{P} \equiv \mathbf{S}$$

The above equality ensures the transformation of the system from an initial point $\mathbf{y}(0) \in S_0$ to a final point $\mathbf{y}(T) \in S_T$ by the vector function $\mathbf{u}(t)$, chosen from U. Thus it is a feasible control. By minimizing J over all feasible sets of $\bar{P}$ and $\bar{\mathbf{S}}$, one can find a feasible solution with the least cost. This is precisely an optimal solution to the continuous-time control problem.

To complete the GP formulation, its subproblem must be described [14, chap. 22]. Here we assume that there are at least $n + 2$ vectors $\mathbf{P}^i$ and/or $\mathbf{S}^i$ available to provide a feasible solution to Eq. (5-105), so that the problem

$$\max_{\mu, \nu} \lambda, \quad \mu, \nu \geqslant 0$$
$$\mathbf{U}_0 \lambda + \mathbf{P}^1 \mu_1 + \mathbf{P}^2 \mu_2 + \cdots + \mathbf{P}^j \mu_j = \mathbf{S}^{j+1}\nu_1 + \cdots + \mathbf{S}^{j+p}\nu_p$$
$$\mu_1 + \mu_2 + \cdots + \mu_j = 1 \quad (5\text{-}106)$$
$$\nu_1 + \nu_2 + \cdots + \nu_p = 1$$

is solvable and has a dual solution vector

$$\bar{\pi}^T = (\pi_0, \pi^T, \pi_{n+1}, \pi_{n+2})$$

where $\pi^T = (\pi_1, \ldots, \pi_n)$. The subproblem is then formulated in two parts:

$$\min_{\mathbf{S} \in S_T} \bar{\pi}^T \begin{bmatrix} \bar{\mathbf{S}} \\ 0 \\ 1 \end{bmatrix} \quad (5\text{-}107)$$

$$\min_{\mathbf{P} \in C} \bar{\pi}^T \begin{bmatrix} \bar{\mathbf{P}} \\ 1 \\ 0 \end{bmatrix} \quad (5\text{-}108)$$

The solution to Eq. (5-107) is dependent on the explicit definition of the set $\bar{S}_T$, the simplest case being the fixed end-point problem, which consists of a single element. In this case, the subproblem (5-107) is trivial and need not be considered. If $\bar{S}_T$ is a convex polyhedral set, then Eq. (5-107) is a linear program that needs to be solved once for each iteration of the master problem.

The subproblem in Eq. (5-108) can be described as

$$\min_{\bar{P} \in C} {}^T\bar{\pi} \begin{bmatrix} \bar{P} \\ 1 \\ 0 \end{bmatrix} = \min_{\bar{P} \in C} \bar{\pi}^T \begin{bmatrix} \int_0^T e^{(T-t)F}\bar{G}\mathbf{u}(t)\,dt \\ 1 \\ 0 \end{bmatrix} \quad (5\text{-}109)$$

Since the requirement $\bar{P} \in \bar{C}$ is equivalent to the requirement $\mathbf{u}(t) \in U$ for all t, Eq. (5-109) becomes

$$\min_{\mathbf{u}(t) \in U} \bar{\pi}^T \begin{bmatrix} \int_0^T e^{(T-t)F}\bar{G}\mathbf{u}(t)\,dt \\ 1 \\ 0 \end{bmatrix}$$

or, since $\bar{\pi}$ does not depend on t,

$$\min_{\mathbf{u}(t) \in U} \left[\int_0^T (\pi_0, \pi^T) e^{(T-t)F} G \mathbf{u}(t)\,dt \right] + \pi_{n+1} \quad (5\text{-}110)$$

The minimum of the integral is attained when the integrand is minimized at every point. Let

$$\boldsymbol{\gamma}(t) = (\pi_0, \pi^T) e^{(T-t)F} \bar{G} \quad (5\text{-}111)$$

be an m-dimensional vector function. Thus the subproblem becomes

$$\text{Find } \mathbf{u}(t) \in U, \quad t \in [0, T]$$
$$\text{So that } \boldsymbol{\gamma}^T(t)\mathbf{u}(t) \text{ is a minimum} \quad (5\text{-}112)$$

From Eq. (5-111), it is obvious that $\boldsymbol{\gamma}(t)$ has the property that each of its components is a member of the class of solutions to an $(n+1)$st-order, homogeneous, constant coefficient, linear differential equation. Since our attention is restricted to those U that are polyhedral sets, Eq. (5-112) becomes

$$\min \boldsymbol{\gamma}^T(t)\mathbf{u}(t)$$
$$A\mathbf{u}(t) \geqslant b, \quad t \in [0, T] \quad (5\text{-}113)$$

(Note that the inequality may be reversed or replaced by an equality without loss of generality.) Thus a solution $\mathbf{u}(t)$ for the subproblem can be obtained by using the parametric LP methods [7, 20].

EXAMPLE

The problem considered is of the minimum fuel type. The performance index is

$$\min_{u(t)} \int_0^3 |u(t)|\,dt$$

The state equations are

$$\dot{y}_1 = y_2$$
$$\dot{y}_2 = u$$

The control constraint is described by

$$|u(t)| \leq 1$$

The initial and terminal states are

$$\mathbf{y}(0) = \begin{bmatrix} 1 \\ 0 \end{bmatrix} \quad \text{and} \quad \mathbf{y}(3) = \begin{bmatrix} 0 \\ 0 \end{bmatrix}$$

Thus

$$F = \begin{bmatrix} 0 & 1 \\ 0 & 0 \end{bmatrix}, \quad G = \begin{bmatrix} 0 \\ 1 \end{bmatrix}, \quad e^{Ft} = \begin{bmatrix} 1 & t \\ 0 & 1 \end{bmatrix}$$

and

$$e^{F(T-t)}G = \begin{bmatrix} 1 & T-t \\ 0 & 1 \end{bmatrix} \begin{bmatrix} 0 \\ 1 \end{bmatrix} = \begin{bmatrix} T-t \\ 1 \end{bmatrix}$$

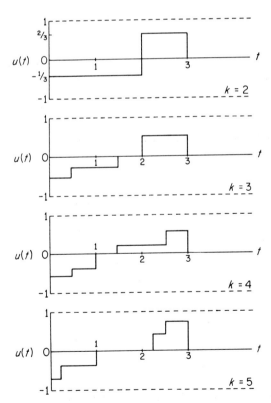

FIG. 5-7(a). $\bar{u}(t)$ vs. t at iterations corresponding to $k = 2, 3, 4, 5$ [7].

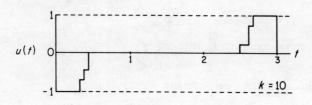
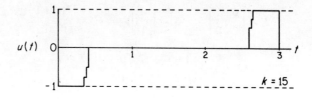
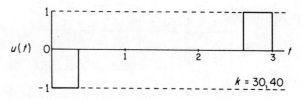

FIG. 5-7(b). $\bar{u}(t)$ vs. t at iterations corresponding to $k = 10, 15, 30, 40$ [7].

Therefore,

$$\mathbf{P} = \int_0^3 \begin{bmatrix} T - t \\ 1 \end{bmatrix} u(t)\, dt$$

and

$$U = \{u \,|\, |u| \leqslant 1\}$$

By the definitions given at the beginning of this section,

$$S_0^F = \left\{ e^{FT} \begin{bmatrix} 1 \\ 0 \end{bmatrix} \right\} = \left\{ \begin{bmatrix} 1 & 3 \\ 0 & 1 \end{bmatrix} \begin{bmatrix} 1 \\ 0 \end{bmatrix} \right\} = \begin{bmatrix} 1 \\ 0 \end{bmatrix}$$

$$S_T = \left\{ \begin{bmatrix} 0 \\ 0 \end{bmatrix} \right\}$$

$$\bar{S}_T = \{S_T - S_0^F\} = \left\{ \begin{bmatrix} -1 \\ 0 \end{bmatrix} \right\}$$

The computational algorithm, described in detail in reference 7, was applied to this problem. The program converged in 40 iterations, using 16-place accuracy, on an IBM 360/67. Figure 5-7 illustrates the control function $\bar{u}^k(t)$ at iterations cor-

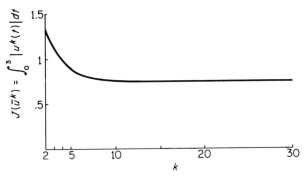

FIG. 5-8. $J(\bar{u}^k)$ vs. k [7].

responding to $k = 2, 3, 4, 5, 10, 15, 30$, and 40. Its cost $J(\bar{u}^k)$ is shown at each iteration in Fig. 5-8.

5.8 APPLICATION IN COMPUTING OPTIMAL CONTROLS FOR A NUCLEAR ROCKET REACTOR [6]

The nuclear rocket reactor to be considered is one which makes use of hydrogen as the propelling medium. Figure 5-9 is a schematic illustration of a nuclear rocket propulsion system. Here, hydrogen is pumped over the reactor core, where heat is transferred to the propellant. The hot hydrogen gas then exits through a nozzle to furnish the thrust. It is assumed that the only variables available to control the nuclear rocket reactor are the hydrogen flow rate, u_2, and the reactivity, u_1.

The differential equations which approximate the nuclear reactor behavior are

$$\dot{Q} = \left(\frac{u_1 - \beta}{l}\right)Q + \lambda C \tag{5-114}$$

$$\dot{C} = \frac{\beta}{l}Q - \lambda C \tag{5-115}$$

where Q is the reactor power level, C is the neutron precursor level, u_1 is the reactivity, u_2 is the hydrogen flow rate, β is the proportion of neutrons which are delayed owing to the precursors group, l is the mean generation time of neutrons, and λ is the average decay constant of the neutron precursors. Equations (5-114) and (5-115) are the classic one-lump, spatially independent neutron kinetics, which are discussed in reference [17].

The heat transfer between the reactor core and the hydrogen is approximated by

$$\dot{T} = \frac{Q}{M_c} - au_2 T \tag{5-116}$$

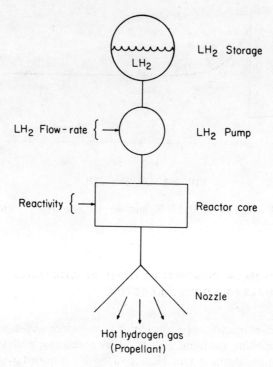

FIG. 5-9. Schematic illustration of the nuclear rocket reactor [6].

where T is the average core temperature, M_c is the effective core mass capacity, and a is a heat-transfer coefficient. Equation (5-116) is a one-lump heat-transfer process.

The specific problem considered is that of reactor startup. Initially the reactor is operating at a relatively low power level. It is desired to transfer the reactor from this low power level to a power level and temperature which corresponds to maximum engine thrust. During such a startup phase, the state and control variables are subject to various constraints that are imposed for mechanical and safety reasons. The constraints which are considered are

$$\begin{aligned} 0 &\leqslant Q \leqslant Q_{\max} \\ 0 &\leqslant C \\ 0 &\leqslant T \leqslant T_{\max} \\ \dot{T} &\leqslant \dot{T}_{\max} \\ -.9\beta &\leqslant u_1 \leqslant .9\beta \\ \dot{u}_1 &\leqslant \dot{u}_{1\,\max} \\ u_{2\,\min} &\leqslant u_2 \leqslant u_{2\,\max} \end{aligned} \qquad (5\text{-}117)$$

The initial boundary conditions $[Q(0), C(0), T(0)]$ are specified by the steady-state low power level. The terminal values of power and temperature $[Q(t_1), T(t_1)]$ are fixed by the maximum thrust criteria. The terminal value of precursor level $[C(t_1)]$ and the terminal time (t_1) itself are allowed to be free, subject, of course, to the restrictions

$$0 \leqslant t_1$$
$$0 \leqslant C(t_1)$$

The criterion of most interest in the operation of the nuclear rocket reactor is the quantity of hydrogen consumed in the process of transferring from one operating condition to another. A small percentage saving in hydrogen consumed can result in a large percentage increase in payload. This is particularly true since not only is a saving effected in the hydrogen proper, but also a saving is effected in the size of the on-board storage tanks. For this reason, the optimality criterion is that the quantity J, defined by

$$J = \int_0^{t_1} u_2(t)\, dt \tag{5-118}$$

has minimum value.

The numerical procedure used to obtain solutions to the optimal nuclear rocket reactor problem makes use of an iterative sequence of LP problems. Figure 5-10 is a schematic illustration of the algorithm. A convergence proof for a very restricted class of problems is presented in reference 18. The first step is to approximate the continuous differential equations and the integral performance criterion by a set of discrete algebraic equations. There are, of course, many methods to obtain these approximations, most of which are patterned after numerical integration procedures. The technique used here for the process differential equations is the second-order approximation

$$Q_{k+1} - Q_k = \frac{\Delta t}{2}\left\{\left(\frac{u_{1_k} - \beta}{l}\right)(Q_{k+1} + Q_k) + \lambda(C_{k+1} + C_k)\right\} \tag{5-119}$$

$$C_{k+1} - C_k = \frac{\Delta t}{2}\left\{\frac{\beta}{l}(Q_{k+1} + Q_k) - \lambda(C_{k+1} + C_k)\right\} \tag{5-120}$$

$$T_{k+1} - T_k = \frac{\Delta t}{2}\left\{\frac{1}{M_c}(Q_{k+1} + Q_k) - au_{2_k}(T_{k+1} + T_k)\right\} \tag{5-121}$$

In obtaining equations (5-119) through (5-121), it is assumed that the controls are stepwise approximations, that is, that the controls u_1 and u_2 are constant over each time interval Δt. Since the approximations are second-order, the truncation error of the Taylor series over any interval t to $t + \Delta t$ (or k to $k + 1$) is proportional to Δt. The interval Δt, while variable, is determined by

$$\Delta t = \frac{(t_1 - t_0)}{N} \tag{5-122}$$

where N is the number of time increments in the interval $t_1 - t_0$.

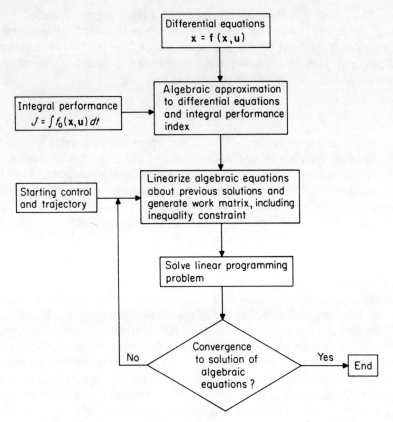

FIG. 5-10. Schematic illustration of the linear-programming algorithm for optimal control problems [6].

Owing to the fact that the controls are piecewise constant over each Δt, the integral performance criterion can be represented algebraically by

$$J = \Delta t \sum_{k=1}^{N} u_{2_k} \qquad (5\text{-}123)$$

Equations (5-119) through (5-121) and (5-123) are nonlinear and thus are not directly suitable for a LP problem (see Chapter 2).

To execute LP procedures, the nonlinear equations (5-119) through (5-121) and (5-123) must be linearized. The linearization is accomplished by retaining the first two terms (the constant and linear terms) of the Taylor series expansion for the nonlinear equations. Since the terminal time is free, Δt is treated as a variable in the Taylor series and in the ensuing LP problems.

Except for the rate constraints, linearization completes the formulation of the LP problem. From a practical standpoint, the rate constraints are easily formulated as linear inequality constraints. That is, the inequality

Sec. 5.8 Nuclear Rocket Reactor 107

constraints
$$u_{1_{k+1}} - u_{1_k} \leqslant \Delta t \, \dot{u}_{1\,\text{max}} \qquad (5\text{-}124)$$
and
$$T_{k+1} - T_k \leqslant \Delta t \, \dot{T}_{\text{max}} \qquad (5\text{-}125)$$

will prevent the reactivity and temperature from increasing too rapidly.

As can be seen from Fig. 5-10, the solution to the nonlinear problem is obtained as the limit of a sequence of LP problems, with each new problem in the sequence resulting from a linearization about the previous solution. Because of the generalized Newtonian character of this process, convergence, if it occurs, will be quadratic.

Most large-scale digital computers have general LP systems available which considerably simplify the programming effort required for the solution of problems in this class. The calculations in this case were performed on an IBM 360/40 digital computer located at the University of New Mexico Computing Center. The Mathematical Programming System/360 (MPS/360) is a general system available for solution of LP problems on the IBM-360 series of computers. Using the source language of MPS/360, only 16 statements were required for the solution of the LP problems associated with the nuclear rocket reactor. This program includes the reinsertion of the previous optimal basis as the starting basis for each new iteration.

The matrix of coefficients and other data associated with the linearized algebraic equations are illustrated in Fig. 5-11. The recursive nature of the

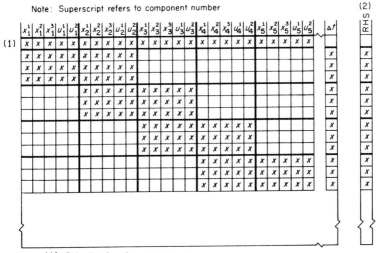

(1) Objective function row
(2) Right-hand side

FIG. 5-11. Illustration of blocked character of work matrix for maximal entries with second-order approximation [6].

algebraic equations and of the inequality rate constraints generates a blocked matrix. The nonzero coefficients of this matrix, together with the coefficients in the right-hand column and the data on bounds for each variable, form the input data for the MPS/360 program. On each iteration these coefficients must be generated in the proper format for MPS/360. Since the matrix is blocked, the data may be generated by a series of DO loops in a FORTRAN program. The complete algorithm shown in Fig. 5-10 was formed by alternately running the FORTRAN and MPS/360 programs. Convergence was indicated when the input data to the MPS/360 program were optimal in accordance with the standard tolerances set in the MPS/360 program.

Computational results for the nuclear rocket reactor have been obtained using the LP algorithm previously described.

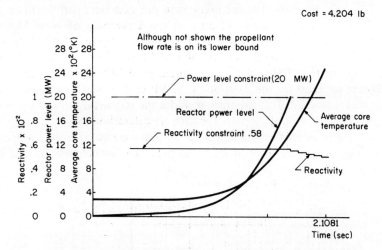

FIG. 5-12. Nuclear rocket reactor trajectories with 20 MW power level constraint [6].

Figure 5-12 shows the trajectories obtained in the absense of the reactivity and temperature-rate constraints. The optimal policy is to insert maximum reactivity until the power-level constraint is reached. The reactivity must then decrease in order to hold the power level on its constraint boundary. The constant power-level condition continues until the terminal-temperature boundary condition is satisfied. During the entire startup process, the propellant flow rate remains on its lower boundary. The process is certainly an absolute minimum, since the integral performance criterion assumes its absolute minimum value under these conditions.

The data illustrated in Fig. 5-13 are similar to that of Fig. 5-12; however, the effective core mass heat capacity was artificially lowered by a factor of 10 so that the temperature constraint was reached prior to the power con-

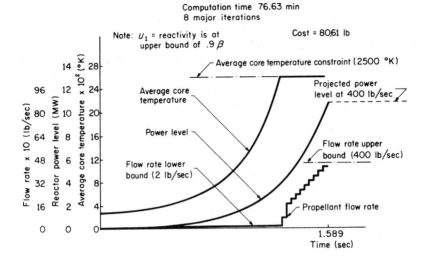

FIG. 5-13. Nuclear rocket reactor trajectories with fast temperature response [6].

straints. In this case, the optimal process is also one of minimum time, but the propellant consumption is much higher than that of Fig. 5-12. The higher propellant consumption results from the fact that the hydrogen flow rate must increase in order to keep the temperature on its boundary during the latter portion of the trajectory. The optimality of the trajectory shown in Fig. 5-13 might be questioned, since there is an alternative policy which can keep the temperature on its boundary. That is, one could allow the propellant flow rate to remain on its lowerbound and use reactivity variations to vary the power level Q so that the temperature differential equation is satisfied. The temperature rate on the temperature boundary is zero, however, so that the power level would have to be constant; i.e.,

$$\dot{T} \equiv 0 = \frac{Q}{M_c} - au_{2\,\min}T_{\max}$$

If the power level remains constant, the terminal power-level boundary condition cannot be satisfied.

REFERENCES

1. C. M. Waespy, "An Application of Linear Programming to Minimum Fuel Optimal Control," Ph.D. Thesis, UCLA, Los Angeles, 1967.
2. A. F. Fath and T. J. Higgins, "Fixed-Time Fuel-Optimal Control of Linear State-Constrained Systems by Use of Linear Programming Techniques," 1968 JACC, pp. 462–467, Ann Arbor, Mich.

3. A. F. Fath, "Approximation to the Time-Optimal Control of Linear State-Constrained Systems," 1968 JACC, pp. 962–969, Ann Arbor, Mich.
4. A. F. Fath, "A Computational Procedure for Fixed-Time Fuel-Optimal Control of Linear State-Constrained Systems," 1968 WESCON, Session 14, Los Angeles, Cal.
5. D. Tabak, "Application of Mathematical Programming in the Design of Optimal Control Systems," Ph.D. Thesis, University of Illinois, Urbana, 1967.
6. H. J. Price and R. R. Mohler, "Computation of Optimal Controls for a Nuclear Rocket Reactor," *IEEE Trans. Nucl. Sci.*, **NS-15**, pp. 65–73, 1968.
7. G. S. Jizmagian, "Generalized Programming Solution of Continuous-Time Linear-System Optimal Control Problems," Ph.D. Thesis, Stanford University, Stanford, Cal., 1968.
8. R. O. Barr, Jr., "Computation of Optimal Controls by Quadratic Programming on Convex Reachable Sets," Ph.D. Thesis, University of Michigan, Ann Arbor, Mich., 1968.
9. L. A. Zadeh and C. A. Desoer, *Linear System Theory*, McGraw-Hill, New York, 1963.
10. B. C. Kuo and D. Tabak, "A Nonlinear Programming Approach to Optimal Control of Discrete Systems with Constrained Sampling Times," 1968 JACC, pp. 918–922, Ann Arbor, Mich.
11. R. A. Volz, "Optimal Control of Discrete Systems with Constrained Sample Times," 1967 JACC, pp. 16–22, Philadelphia, Pa.
12. W. H. Clohessy and R. S. Wiltshire, "Terminal System for Satellite Rendezvous," *J. Aerospace Sci.*, **27**, pp. 653–658, Sept. 1960.
13. "LP/90 Usage Manual," 2nd ed., CEIR, Inc., Washington, D.C., July 31, 1963.
14. G. B. Dantzig, *Linear Programming and Extensions*, Princeton University Press, Princeton, N.J., 1963.
15. G. B. Dantzig, "Linear Control Processes and Mathematical Programming," *J. SIAM Control*, **4**, pp. 56–60, 1966.
16. M. Van Slyke, "Mathematical Programming and Optimal Control," Ph.D. Thesis, University of California, Berkeley, 1965.
17. M. Ash, *Nuclear Reactor Kinetics*, McGraw-Hill, New York, 1965.
18. J. B. Rosen, "Iterative Solution of Nonlinear Control Problems," *J. SIAM Control*, **4**, pp. 223–224, 1966.
19. G. S. Jizmagian, "Generalized Programming Solution of Optimal Control Problems," in *Computing Methods in Optimization Problems*, vol. 2 (L. A. Zadeh, L. W. Neustadt, and A. V. Balakrishnan, eds.), Academic Press, New York, 1969, pp. 165–176.
20. G. S. Jizmagian, "An Algorithm for Parametric Linear Programming with Nonlinear Functions of the Parameter," 10th Meeting of The Institute of Management Science, Atlanta, Georgia, Oct. 1969.

6

NONLINEAR CONTINUOUS-TIME SYSTEMS

6.1 INTRODUCTION

The general formulation of an optimal-control problem as a mathematical-programming problem for a continuous-time nonlinear system has been outlined in Chapter 4 [Eq. (4-26)]. Some computational examples illustrating the implementation of this formulation will be presented in this chapter. In addition, other formulations for applying MP techniques to the numerical solution of optimal-control problems for nonlinear continuous-time systems will be presented.

6.2 XENON POISONING CONTROL OF NUCLEAR REACTORS

As an illustrative example of a MP solution of the optimal-control problem formulated in Eq. (4-26), the problem of optimal shutdown control of nuclear reactors under xenon poisoning is considered [1–6], described as follows:

During the operation of a nuclear reactor several kinds of nuclides with very high neutron absorption probabilities are generated as fission fragments. The strongest neutron absorbers are the xenon (Xe) and iodine (I). The strongest is Xe, which is obtained as a result of radioactive decay of I. When the reactor is shut down and the neutron flux is reduced practically to zero, more Xe is generated and then lost owing to absorption, and the Xe obtains a peak value. Since the half-time of Xe decay is 9.2 hr, it takes quite a long time until the quantity of Xe decreases enough to permit practical operation of the reactor. Mathematically, the process can be described in the following way. If we treat the concentrations of Xe and I as state variables and neutron flux as the control variable, then the basic state equations describing the process are [3]

$$\dot{x} = -(w + r_0 u)x + g_0 y + g_2 u$$
$$\dot{y} = -y + u \qquad (6\text{-}1)$$

where

$g_1 = \gamma_1(w + r_0)$
$g_2 = \gamma_2(w + r_0)$
$x = $ Scaled, dimensionless Xe concentration
$y = $ Scaled, dimensionless I concentration
$u = $ Scaled, dimensionless flux $= \phi/\phi_0$
$\gamma_1 = \gamma_I/(\gamma_x + \gamma_I)$
$\gamma_2 = \gamma_x/(\gamma_x + \gamma_I)$
$r_0 = \dfrac{\sigma \phi_0}{\gamma_1}$
$w = \dfrac{\lambda_2}{\lambda_1}$
$\gamma_x = $ Relative yield of Xe produced from fission
$\gamma_I = $ Relative yield of I produced from fission
$\phi = $ Thermal neutron flux
$\phi_0 = $ Basic flux reference unit
$\lambda_1 = $ Decay constant of I
$\lambda_2 = $ Decay constant of Xe
$\sigma = $ Microscopic absorption cross section of Xe for thermal neutrons

As we can see, the state equations in Eq. (6-1) are nonlinear because of the term ux appearing in the first equation. In this case, we work with a time scale in units of $\lambda_1^{-1} = 9.58$ hr. The scaling of all the variables is performed in such a way that the initial conditions are

$$x(0) = y(0) = 1$$

and

$$\dot{x}(0) = \dot{y}(0) = 0$$

At the moment when shutdown is completed, i.e., when the flux u is brought to zero, the equations in Eq. (6-1) become linear:

$$\dot{x} = -wx + g_1 y$$
$$\dot{y} = -y \qquad (6\text{-}2)$$

The solution of Eq. (6-2) can easily be obtained.

If the shutdown occurred at time $t = T$, when x and y have the values $x(T)$ and $y(T)$, respectively, the solution of Eq. (6-2) at a time $t > T$ would be [3]

$$x(t) = x(T) \exp[-w(t-T)] + y(T) \left\{ \frac{g_1}{1-w} \exp[-w(t-T)] - \exp[-(t-T)] \right\} \quad (6\text{-}3)$$

By direct differentiation, we can obtain the value of the Xe peak, i.e., $x(t)$ maximum:

$$x_p = Gy(T) \left[1 + F \frac{x(T)}{y(T)} \right]^{g_w} \quad (6\text{-}4)$$

where

$$G = g_1 w^{w/(1-w)}$$

$$F = \frac{1-w}{g_1}$$

$$g_w = \frac{1}{1-w}$$

The lower the value of x_p, the sooner the reactor may be restarted. This factor is of particular importance in vehicular nuclear reactors where fast restarts may be critical in many occasions.

There are several ways of posing the control problem. Two approaches are described in the following.

1. The Minimax Problem [3]

For a given shutdown time T, what should be the control law $u(t)$ during the shutdown period, $0 < t \leqslant T$, so that the xenon peak, which is a function of the final state Eq. (6-4), is minimized? This is obviously a terminal-control problem. It has been attacked previously by Ash [3] using dynamic programming [7]. The following state and control constraints are posed:

$$x \leqslant x_{max} \quad (6\text{-}5)$$

$$u \leqslant u_{max} \quad (6\text{-}6)$$

It is obvious that because of physical reasons, all the state and control variables treated here are nonnegative.

2. The Minimum-Time Problem [4, 5]

In this case, the final state at $t = T$ is such that the x trajectory from that point attains a peak of x: $x_p = x_{max}$. The set of possible terminal points form a locus in the yx plane, whose equation is given by [4]

$$x = g(y) = \left[\frac{x_{max}}{(x_{max}w)^w} + \frac{(x_{max}w)^{1-w}}{1-w} \right] y^w - \frac{y}{1-w} \quad (6\text{-}7)$$

The problem would then be to reach the target curve $g(y)$ in minimum time. It has been shown by Roberts and Smith [5] that the two problems described in 1 and 2 are equivalent.

In this work, before applying any modern method of optimization, the problem was studied through a direct classic approach. If we want to minimize the xenon peak x_p, we should look at the variables it depends on. As we can see from Eq. (6-4), x_p depends on the terminal state, $x(T)$ and $y(T)$ only. Theoretically, there is an infinite number of shutdown histories and control laws by which a certain terminal state can be obtained; however, x_p would still be the same. Therefore, before considering what the optimal law during the shutdown period should be, we should ask what the terminal state $x(T)$, $y(T)$ should be so that x_p is minimized. By direct partial differentiation of Eq. (6-4), for any value of $x(T)$, x_p will be minimal if the following ratio is satisfied:

$$\frac{x(T)}{y(T)} = \frac{1}{F(g_w - 1)} \tag{6-8}$$

By choosing Eq. (6-8) as the target curve in the yx plane, we can be assured of a minimum xenon peak x_p no matter what final value of xenon concentration x is reached. Now we can pose a modified optimal-control problem as follows:

Given an initial state, $[x(0), y(0)]$, find an optimal-control law $u(t)$ so that the target set, described by Eq. (6-8), is reached in minimum time.

Using the formulation of Chapter 4, the state equations in Eq. (6-1) are now rewritten in discrete-time form and will serve as equality constraints in the nonlinear-programming problem:

$$\begin{aligned} x_{i+1} - x_i &= -(w + r_0 u_i) x_i T_{i+1} + g_1 y_i T_{i+1} + g_2 u_i T_{i+1} \\ y_{i+1} - y_i &= -y_i T_{i+1} + u_i T_{i+1}, \quad i = 0, 1, \ldots, N-1 \end{aligned} \tag{6-9}$$

The performance index will now be: minimize

$$J = \sum_{i=1}^{N} T_i \tag{6-10}$$

subject to Eq. (6-9) and the following inequality constraints:

$$x_i \leqslant x_{\max}, \quad i = 1, \ldots, N \tag{6-11}$$

$$u_i \leqslant u_{\max}, \quad i = 0, \ldots, N-1 \tag{6-12}$$

Two numerical examples are solved in the following; one corresponds to the data used in reference 3, and the other is close to those in reference 4.

EXAMPLE 1[†]

The data used:
$$x(0) = y(0) = 1$$
$$u_{\max} = 2$$

[†]From reference 3.

$$x_{max} = 4.8$$
$$\gamma_x = .003$$
$$\gamma_y = .056$$
$$\lambda_1 = 2.9 \times 10^{-5} \text{ sec}^{-1} \text{ (I)}$$
$$\lambda_2 = 2.1 \times 10^{-5} \text{ sec}^{-1} \text{ (Xe)}$$
$$\sigma = 3.5 \times 10^6 \text{ barns} = 3.5 \times 10^{-18} \text{ cm}^2$$
$$\phi_0 = 1.66 \times 10^{14} \text{ n/cm}^2 \text{ sec}$$

The computed values:
$$r_0 = 20$$
$$w = .724$$
$$\gamma_1 = .949$$
$$\gamma_2 = .0508$$
$$g_1 = 19.67$$
$$g_2 = 1.054$$
$$g_w = 3.625$$
$$F = .014$$
$$G = 8.4$$

The following results were obtained in the NLP run for $N = 12$:

i	T_i (min)	t_i (min)	$u_{i-1} \times 10^{-3}$	x_i (Xe)	y_i (I)
1	.032	.032	4.67	1.001	1.000
2	.032	.064	4.67	1.002	1.000
3	.032	.096	4.67	1.003	1.000
4	.032	.128	4.67	1.004	1.000
5	.032	.160	4.67	1.005	1.000
6	.032	.192	4.67	1.006	1.000
7	.032	.224	4.67	1.007	1.000
8	115.035	115.259	.07	4.796	.799
9	.032	115.291	2.14	4.797	.799
10	.032	115.323	2.14	4.797	.799
11	.032	115.355	2.14	4.798	.799
12	533.921	649.276	129.05	4.800	.177

As we can see,
$$T_{min} = 649.276 \text{ min} = 10.8 \text{ hr}$$

The T_i's actually appearing in the computer program are the T_i's in the table divided by the scaling factor of 574.8 min.

EXAMPLE 2†

The data used:
$$x(0) = y(0) = 1$$
$$u_{max} = 1$$

---
†From reference 4.

116 *Nonlinear Continuous-Time Systems*   Ch. 6

$$x_{max} = 5.0$$
$$\gamma_x = .002$$
$$\gamma_y = .061$$
$$\lambda_1 = 2.9 \times 10^{-5} \text{ sec}^{-1}$$
$$\lambda_2 = 2.1 \times 10^{-5} \text{ sec}^{-1}$$
$$\sigma = 2.7 \times 10^6 \text{ barns} = 2.7 \times 10^{-18} \text{ cm}^2$$
$$\phi_0 = 10^{14} \text{ n/cm}^2 \text{ sec}$$

The computed values (different from those in the previous example):

$$r_0 = 9.47$$
$$\gamma_1 = .968$$
$$\gamma_2 = .0317$$
$$g_1 = 9.87$$
$$g_2 = .324$$
$$F = .02795$$
$$G = 4.21$$

The following results were obtained for $N = 12$:

i	T_i (min)	t_i (min)	$u_{i-1} \times 10^{-3}$	x_i (Xe)	y_i (I)
1	117.000	117.000	.47	2.86	.796
2	.272	117.272	12.00	2.86	.796
3	.270	117.542	12.00	2.86	.795
4	.266	117.808	12.00	2.86	.795
5	.271	118.079	12.00	2.87	.794
6	.270	118.349	12.00	2.87	.794
7	.270	118.619	12.00	2.88	.793
8	.272	118.891	12.00	2.88	.793
9	.273	119.164	12.00	2.88	.792
10	.275	119.439	12.00	2.88	.792
11	.274	119.713	12.00	2.89	.791
12	343.000	462.713	80.78	5.00	.367

The minimum time obtained is

$$T_{min} = 462.713 \text{ min} = 7.7 \text{ hr}$$

To check for possible discretization errors, the same example was run for $N = 20$. The final results turned out to be about the same, with the minimum shutdown time being

$$T_{min} = 463 \text{ min}$$

Considering such a small difference, and a small relative difference,

$$\frac{T_{min}^{20} - T_{min}^{12}}{T_{min}^{20}} = \frac{.7}{463} = .0015 = .15\%$$

we can argue that the discretization error is tolerable for practical purposes.

Sec. 6.3   Higher-Order Approximation; A Trajectory Optimization Problem   **117**

These examples were run on the IBM 7094 Computer utilizing the SUMT method. Looking again at the discretized state equations in Eq. (6-9), we can see that the simplest form of discretization was used. This discretization is a first-order approximation only; it constitutes the first two terms of a Taylor expansion, up to the first-order derivative. To obtain better precision in the discretization of the nonlinear continuous-time state equations, we have to use approximations of higher order. An example illustrating this point is presented in Sec. 6.3.

6.3 HIGHER-ORDER APPROXIMATION; A TRAJECTORY OPTIMIZATION PROBLEM

The use of a third-order differencing scheme has been proposed by Pierson [8, 9]. The method of discretization follows the Adams interpolation difference equations [10]. Starting with a set of state equations as in Eq. (4-1), we subdivide the time interval $t_0 \leqslant t \leqslant t_f$ into N equal intervals. The difference equations are then formulated as follows:

$$\begin{aligned}
\mathbf{y}(1) - \mathbf{y}(0) &- 12b[\mathbf{f}(0) + \mathbf{f}(1)] = 0 \\
\mathbf{y}(2) - \mathbf{y}(1) &- 2b[-\mathbf{f}(0) + 8\mathbf{f}(1) + 5\mathbf{f}(2)] = 0 \\
\mathbf{y}(3) - \mathbf{y}(2) &- b[\mathbf{f}(0) - 5\mathbf{f}(1) + 19\mathbf{f}(2) + 9\mathbf{f}(3)] = 0 \\
&\vdots \\
\mathbf{y}(N) - \mathbf{y}(N-1) &- b[\mathbf{f}(N-3) - 5\mathbf{f}(N-2) + 19\mathbf{f}(N-1) + 9\mathbf{f}(N)] = 0
\end{aligned} \qquad (6\text{-}13)$$

where

$$\mathbf{f}(i) = \mathbf{f}[\mathbf{y}(i), \mathbf{u}(i), i] \qquad (6\text{-}14)$$

$$b = \frac{t_f}{24N} \qquad (6\text{-}15)$$

The value of b in Eq. (6-15) has proved to be quite satisfactory in previous computational experience [8, 9].

The numerical algorithm used in this case is similar to the SUMT algorithm (see Chapter 3) in the sense that it involves the solution of a sequence of unconstrained problems. However, in this algorithm a different penalty method is used. The method using the Heaviside unit function was originally proposed by Courant [11, 12].

The modified performance index used in this algorithm is

$$J^* = J + K_v \left[\sum_i h_i^2 + \sum_j H(-g_j)(wg_j)^4 \right] \qquad (6\text{-}16)$$

where J = Original performance index
h_i = Equality constraints, $h_i = 0$
g_j = Inequality constraints, $g_j \geqslant 0$
H = Heaviside unit step function:

$$H(z) = \begin{cases} 1, & z > 0 \\ 0, & z \leqslant 0 \end{cases} \tag{6-17}$$

$K_v = C^{k-1} K_1 ;\ C > 1;\ K_1 > 0$
$k = 1, 2, \ldots$ (K_v plays the role of r_i in the SUMT algorithm, as explained in Chapter 3)
w = Arbitrary weighting constant

The unconstrained minimization method applied in this case belongs to the class of conjugate-gradient methods (see Sec. 3.4). To be more specific, the method of Fletcher and Reeves [13] was used. Using this method, the new direction vector $\mathbf{r}_i$ is chosen according to the following formula:

$$\mathbf{r}_i = -\nabla f_{i+1} + \mathbf{r}_{i-1} \frac{(\nabla f_i^T \nabla f_i)}{(\nabla f_{i-1}^T \nabla f_{i-1})} \tag{6-18}$$

The iterations are restarted after every $M + 1$ steps, where M is the total number of variables in the minimization problem. This is based mainly on computational experience in using the method [8]. Three different search methods for the determination of the step size d_i, at each iteration, have been tried: a modified Fibonacci search [14], a modified Newton method [15], and a cubic interpolation suggested by Davidon [16]. The last has proved to be considerably more efficient than either of the first two [8].

EXAMPLE 3

A Simple Rocket Problem†

The problem considered could be stated as follows:
Assuming constant gravitational acceleration g_0, vacuum flight, and constant rocket thrust acceleration A, find the thrust direction time history required for the rocket to attain a given altitude, Y_f, at a given time, t_f, with zero vertical speed, V_y, and maximum horizontal speed, V_x. The terminal range is not specified.
The performance index for this problem is

$$\text{Maximize } V_x(t_f)$$

or

$$\text{Minimize } J = -V_x(t_f) \tag{6-19}$$

---

†From reference 8.

Sec. 6.3 Higher-Order Approximation; A Trajectory Optimization Problem

The state vector for this problem is

$$\mathbf{y} = \begin{bmatrix} X \\ Y \\ V_x \\ V_y \end{bmatrix} \begin{matrix} \text{horizontal range} \\ \text{altitude} \\ \text{horizontal velocity component} \\ \text{vertical velocity component} \end{matrix}$$

The state equations for this system are

$$\begin{aligned} \dot{y}_1 &= y_3 \\ \dot{y}_2 &= y_4 \\ \dot{y}_3 &= A \cos u \\ \dot{y}_4 &= A \sin u - g_0 \end{aligned} \qquad (6\text{-}20)$$

The control variable u is the inclination of the thrust vector to the horizontal.
The initial condition is

$$y_i(0) = 0, \; i = 1, \ldots, 4 \qquad (6\text{-}21)$$

and the terminal conditions are

$$\begin{aligned} y_2(t_f) &= Y_F = 10^5 \text{ ft} \\ y_4(t_f) &= 0 \\ t_f &= 100 \text{ sec} \end{aligned} \qquad (6\text{-}22)$$

Other pertinent data were

$$g_0 = 32 \text{ ft/sec}^2$$
$$A/g_0 = 2$$

The problem was solved on the IBM 7090 computer, and the results may be summarized as follows:

N	K_v	Computing time (sec)	Iterations	Final V_x (ft/sec)
5	1,000		123	3688.5
5	4,000		19	3632.0
5	16,000		20	3113.4
5	64,000	10	18	3608.9
10	64,000		398	3517.6
10	256,000	56	37	3514.7
20	256,000	77	95	3500.0
40	256,000	342	654	3502.6

As we can see, increasing the number of sampling intervals beyond $N = 20$ does not change the final value of V_x considerably. The complete solution for $N = 20$ is shown in Fig. 6-1.

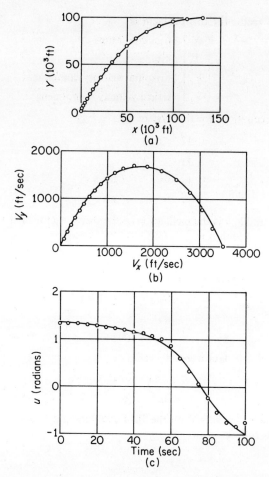

FIG. 6-1. Results for the simple rocket problem [8]: (a) optimal trajectory; (b) optimal trajectory hodograph; (c) optimal control time history.

6.4 COMPUTER ADJUSTMENT OF PARAMETERS OF NONLINEAR CONTROL SYSTEMS

A control system, or process, may often by required to work under changing performance specifications. Sometimes it is more important to work with the highest possible gain, while in another stage of the process, minimum rise time may be of crucial importance. In all cases, one has to maintain the stability of the system. Operating the system with varying optimal performance criteria may be achieved by changing the adjustable parameters of the system in an appropriate manner. One of the best ways of controlling the system under the mentioned conditions is by using direct or indirect digital computer control.

An algorithm combining the system's dynamics, stability criteria, and the performance index, which calculates the optimal system parameters, is programmed. As the performance index changes, it is appropriately put into the computer, and the optimal parameters are recalculated. The output values may then be transmitted directly or indirectly into the system.

A fast algorithm accomplishing the mentioned task for nonlinear control systems has been worked out [17]. The algorithm consists of a combination of Popov's stability criterion for nonlinear systems [18], and of an NLP technique. The algorithm has been simulated on the IBM 360/91 System.

The system under consideration is represented schematically in Fig. 6-2(a). It consists of an nth-order linear system, whose transfer function is $G(s, \mathbf{p})$. $\mathbf{p}$ is the parameters vector, whose components are the adjustable parameters of the system. $G(s, \mathbf{p})$ may in general include the original plant as well as the controller. The system includes a single nonlinearity, which is assumed separable from the linear part. A general form of the nonlinearity transfer function is sketched in Fig. 6-2(b). The input to the nonlinearity is the error signal e, and the output, $u = n(e)$, serves as the input to the linear part of the system. The following assumptions [18] concerning the nonlinearity are added:

1. $n(e)$ is defined and continuous for all values of e.
2. $n(0) = 0$ and $en(e) > 0$ for all $e \neq 0$.

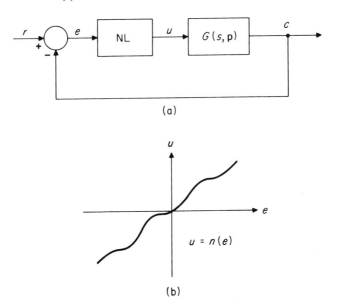

FIG. 6-2. The control system [17]: (a) system block diagram; (b) the nonlinearity.

Although the nonlinearity is restricted in its nature, a wide class of nonlinearities, existing in practice, fits into this category, or could be represented by it approximately.

In the frequency domain, the linear subsystem transfer function may be represented as

$$G(j\omega, \mathbf{p}) = G_1(\omega, \mathbf{p}) + jG_2(\omega, \mathbf{p}) \qquad (6\text{-}23)$$

It will be assumed that all of the poles and zeros of the linear subsystem $G(s, \mathbf{p})$ are in the left-hand plane.

The theorem of Popov [18] states that the system under consideration will be absolutely stable if for

$$0 \leqslant \frac{n(e)}{e} \leqslant K \qquad (\text{or } 0 \leqslant u/e \leqslant K)$$

there exists a nonnegative real q such that

$$G_1(\omega, \mathbf{p}) - q\omega G_2(\omega, \mathbf{p}) + \frac{1}{K} > 0 \qquad (6\text{-}24)$$

K actually represents the maximum allowable instant gain (ratio between output and input) of the nonlinearity. As will be seen from subsequent examples, we can deduce from the inequality in Eq. (6-24) a set of inequality constraints which are functions of $\mathbf{p}$, K, and q:

$$g_j(\mathbf{p}, K, q) \geqslant 0, \qquad q \geqslant 0 \qquad (6\text{-}25)$$

where $j = 1, \ldots, M$. The total number M of constraints in Eq. (6-25) derived from the stability condition in Eq. (6-24) depends on the order and the particular structure of the linear subsystem.

At different times we can impose on the system certain performance indices of the general form

$$J = f(\mathbf{p}, K) \qquad (6\text{-}26)$$

An optimization problem would then be to minimize (or maximize) J, subject to constraints in Eq. (6-25), as well as to any other additional constraints that may be imposed on the system because of various practical considerations:

$$g_i(\mathbf{p}, K) \geqslant 0, \qquad i = 1, \ldots, I_g$$

and/or

$$h_i(\mathbf{p}, K) = 0, \qquad i = 1, \ldots, I_h$$

This formulation constitutes a NLP problem.

EXAMPLE 4

A Third-Order System [17]

The general form of the third-order system was chosen to be

$$G(s, \mathbf{p}) = \frac{1}{(s + a)(s^2 + 2\delta\omega_n s + \omega_n^2)} \qquad (6\text{-}29)$$
$$a, \delta, \omega_n \geqslant 0$$

Sec. 6.4 Computer Adjustment of Parameters of Nonlinear Control Systems

In this case the adjustable parameters are a, δ, and ω_n, and hence the parameter vector is

$$\mathbf{p} = \begin{bmatrix} a \\ \delta \\ \omega_n \end{bmatrix}$$

$$G(j\omega, \mathbf{p}) = G_1(\omega, \mathbf{p}) + jG_2(\omega, \mathbf{p})$$

Substituting $s = j\omega$ into Eq. (6-29), we obtain

$$G_1(\omega, \mathbf{p}) = \frac{a(\omega_n^2 - \omega^2) - 2\delta_n \omega^2}{(a^2 + \omega^2)[(\omega_n^2 - \omega^2)^2 + 4\delta^2 \omega_n^2 \omega^2]} \tag{6-30}$$

$$G_2(\omega, \mathbf{p}) = -\frac{\omega(\omega_n^2 - \omega^2) + 2\delta a \omega_n \omega}{(a^2 + \omega^2)[(\omega_n^2 - \omega^2)^2 + 4\delta^2 \omega_n^2 \omega^2]} \tag{6-31}$$

Substituting Eqs. (6-30) and (6-31) into Eq. (6-24) and introducing

$$x = \frac{1}{K}$$

we obtain

$$x\omega^6 + [x(a^2 - 2\omega_n^2 + 4\delta^2 \omega_n^2) - q]\omega^4 + [x(4\delta^2 a^2 \omega_n^2 - 2a^2 \omega_n^2 + \omega_n^4) \\ + 2q\delta a \omega_n + q\omega_n^2 - 2\delta \omega_n - a]\omega^2 + (xa^2 \omega_n^4 + a\omega_n^2) > 0 \tag{6-32}$$

Since both x and q are nonnegative, and since Eq. (6-32) contains terms of ω of even order only, the inequality in Eq. (6-32) will be satisfied for all ω if the coefficients of ω^4 and ω^2 are nonnegative; i.e.,

$$x(a^2 - 2\omega_n^2 + 4\delta^2 \omega_n^2) - q \geqslant 0 \tag{6-33}$$

$$x(4\delta^2 a^2 \omega_n^2 - 2a^2 \omega_n^2 + \omega_n^4) + 2q\delta a \omega_n + q\omega_n^2 - 2\delta \omega_n - a \geqslant 0 \tag{6-34}$$

The constraints in Eqs. (6-33) and (6-34) constitute the stability constraints of the general form of Eq. (6-25) for the third-order system. In addition to the stability constraints in Eqs. (6-33) and (6-34), the following constraints have been imposed in this example:

$$\begin{aligned} \delta_{\min} &\leqslant \delta \leqslant \delta_{\max} \\ a_{\min} &\leqslant a \leqslant a_{\max} \\ \omega_n &\leqslant \omega_{n\,\max} \\ q &\leqslant q_{\max} \end{aligned} \tag{6-35}$$

Although the constraints in Eq. (6-35) are linear and very simple, the constraints in Eqs. (6-33) and (6-34) are nonlinear; Eq. (6-33) is fifth order and Eq. (6-34) is seventh order.

The performance index for this example was chosen according to the following considerations. On many occasions, we want to adjust the system parameters so that we can work with the maximum forward gain (which in this case happens to be K) and still retain the stability of the system. On the other hand, we may want to have the shortest possible rise time, which would require working with a low damping ratio δ. The requirement of maximizing K (or minimizing $x = 1/K$) would require a higher value of δ. The two requirements drive the optimization problem in opposite directions. In the combined performance index, we should assign an appropriate weighting factor to show the relative importance of both requirements. Since both x and δ are nonnegative, the performance index could be chosen to be linear instead of quadratic†:

$$\text{Minimize } J = cx + \delta \tag{6-36}$$

The following limiting constraints were imposed in addition to the stability constraints in Eqs. (6-33) and (6-34):

$$\begin{aligned} .5 &\leqslant \delta \leqslant .707 \\ .1 &\leqslant a \leqslant 1 \\ \omega &\leqslant 3 \text{ rad/sec} \\ q &\leqslant 10 \end{aligned} \tag{6-37}$$

The variables were assigned as follows:

Control-problem variables	MP variables
x	$x(1)$
a	$x(2)$
δ	$x(3)$
ω_n	$x(4)$
q	$x(5)$

Substituting the MP notation into Eqs. (6-33), (6-34), (6-36), and (6-37), we obtain the following NLP problem:

$$\min \{cx(1) + x(3) \mid x(3) - \delta_{\min} \geqslant 0;\ x(2) - a_{\min} \geqslant 0;\ a_{\max} - x(2) \geqslant 0;$$
$$\omega_{n\max} - x(4) \geqslant 0;\ x(1)\{x^2(2) - [2 - 4x^2(3)]x^2(4)\} - x(5) \geqslant 0;$$
$$x(1)\{[4x^2(3) - 2]x^2(2)x^2(4) + x^4(4)\} + 2x(2)x(3)x(4)x(5) + x(5)x^2(4)$$
$$- 2x(3)x(4) - x(2) \geqslant 0;\ \delta_{\max} - x(3) \geq 0;\ q_{\max} - x(5) \geqslant 0\}$$

The NLP problem has five variables and eight inequality constraints. Since two of the constraints are nonlinear, the SUMT program (Chapter 3) is a good candidate for implementation in this case. To make the problem amenable to direct SUMT computation, we have to provide the program with four auxiliary subroutines:

1. READIN. This subroutine reads in the data of the problem to be solved and sets up various constants to be used later in the program.
2. RESTNT. This subroutine calculates the values of the objective function and of the constraints for a given set of the x variables and places them in the VAL location.

---

†As a result of maximizing K, we may obtain the value of K at the limits of stability. In this case the result should be indicative only to the designer; it does not mean that the system should be actually working with this maximum value.

3. GRAD1. This subroutine calculates the first-order derivatives vector (the gradient) for the objective function and for each constraint. The result is stored in array DEL.
4. MATRIX. This subroutine calculates the second-order derivatives matrix for the objective function and for each constraint. The result is stored in the two-dimensional array A. Since this matrix is symmetrical, only the upper half of the matrix and its diagonal are calculated.

The subroutines RESTNT, GRAD1, and MATRIX are called for the objective function and for each constraint separately. The basic features of SUMT have been described in Sec. 3.3. A detailed program-user's write-up is available on SHARE, No. 3189. The coding of the four auxiliary subroutines used in this problem is given in Table 6-1.

Table 6-1

```
        SUBROUTINE READIN
C       POPOV-NLP COMBINATION
C       THIRD ORDER SYSTEM
        COMMON/BLK/DM,AMN,AMX,WNM,DMX,QM,CO
        READ(5,101)DM,AMN,AMX,WNM,DMX,QM
  101   FORMAT(6E13.4)
        READ(5,105) CO
  105   FORMAT(E20.8)
        RETURN
        END

        SUBROUTINE RESTNT(IN,VAL)
C       POPOV-NLP COMBINATION
C       THIRD ORDER SYSTEM
        COMMON/BLK/DM,AMN,AMX,WNM.DMX,QM,CO
        COMMON/SHARE/X(100),DEL(100),A(100,100),N,M,MN,NP1,NM1
        IF(IN)5,5,20
    5   VAL=CO*X(1)+X(3)
        RETURN
   20   GO TO (21,22,23,24,25,26,27,28),IN
   21   VAL=X(3)-DM
        RETURN
   22   VAL=X(2)-AMN
        RETURN
   23   VAL=AMX-X(2)
        RETURN
   24   VAL=WNM-X(4)
        RETURN
```

Table 6-1—*Cont.*

```
25      VAL=X(1)*(X(2)**2-(2.-4.*X(3)**2)*X(4)**2)-X(5)
        RETURN
26      VAL=X(1)*((4.*X(3)**2-2.)*(X(2)**2)*(X(4)**2)+X(4)**4)+
        12.*X(5)*X(3)*X(2)*X(4)+X(5)*X(4)**2-2.*X(3)*X(4)-X(2)
        RETURN
27      VAL=DMX-X(3)
        RETURN
28      VAL=QM-X(5)
        RETURN
        END


        SUBROUTINE GRAD1(IN)
C       POPOV-NLP COMBINATION
C       THIRD ORDER SYSTEM
        COMMON/BLK/DM,AMN,AMX,WNM,DMX,QM,CO
        COMMON/SHARE/X(100),DEL(100),A(100,100),N,M,MN,NP1,NM1
        DO 10 I=1,N
10      DEL(I)=0.
        IF(IN)5,5,20
5       DEL(1)=CO
        DEL(3)=1.
        RETURN
20      GO TO (21,22,23,24,25,26,27,28),IN
21      DEL(3)=1.
        RETURN
22      DEL(2)=1.
        RETURN
23      DEL(2)=-1.
        RETURN
24      DEL(4)=-1.
        RETURN
25      DEL(1)=X(2)2**-(2.-4.*X(3)**2)*X(4)**2
        DEL(2)=2.*X(1)*X(2)
        DEL(3)=8.*X(1)*X(3)*X(4)**2
        DEL(4)=-2.*X(1)*(2.-4.*X(3)**2)*X(4)
        DEL(5)=-1.
        RETURN
26      DEL(1)=(4.*X(3)**2-2.)*(X(2)**2)*(X(4)**2)+X(4)**4
        DEL(2)=2.*X(1)*(4.*X(3)**2-2.)*X(2)*(X(4)**2)+2.*X(5)*X(3)*X(4)
       1-1.
```

Table 6-1—*Cont.*

```
      DEL(3)=8.*X(1)*X(3)*(X(2)**2)*X(4)**2+2.*X(5)*X(2)*X(4)
     1-2.*X(4)
      DEL(4)=2.*X(1)*((4.*X(3)**2-2.)*(X(2)**2)*X(4)+2.*X(4)**3)+
     12.*X(5)*X(3)*X(2)+2.*X(5)*X(4)-2.*X(3)
      DEL(5)=2.*X(2)*X(3)*X(4)+X(4)**2
      RETURN
27    DEL(3)=-1.
      RETURN
28    DEL(5)=-1.
      RETURN
      END


      SUBROUTINE MATRIX (IN)
C     POPOV-NLP COMBINATION
C     THIRD ORDER SYSTEM
      COMMON/BLK/DM,AMN,WNM,DMX,QM,CO
      COMMON/SHARE/X(100),DEL(100),A(100,100),N,M,MN,NP1,NM1
      IF(IN)5,5,20
5     RETURN
20    GO TO (21,21,21,21,25,26,21,21),IN
21    RETURN
25    A(1,2)=2.*X(2)
      A(1,3)=8.*X(3)*X(4)**2
      A(1,4)=-2.*(2.-4.*X(3)**2)*X(4)
      A(2,2)=2.*X(1)
      A(3,3)=8.*X(1)*X(4)**2
      A(3,4)=16.*X(1)*X(3)*X(4)
      A(4,4)=-4.*X(1)*(1.-2.*X(3)**2)
      RETURN
26    A(1,2)=2.*(4.*X(3)**2-2.)*X(2)*(X(4)**2)
      A(1,3)=8.*X(3)*(X(2)**2)*(X(4)**2)
      A(1,4)=2.*(4.*X(3)**2-2.)*(X(2)**2)*X(4)+4.*X(4)**3
      A(2,2)=2.*X(1)*(4.*X(3)**2-2.)*(X(4)**2)
      A(2,3)=16.*X(1)*X(3)*X(2)*(X(4)**2)+2.*X(5)*X(4)
      A(2,4)=4.*X(1)*(4.*X(3)**2-2.)*X(2)*X(4)+2.*X(5)*X(3)
      A(2,5)=2.*X(3)*X(4)
      A(3,3)=8.*X(1)*(X(2)**2)*(X(4)**2)
      A(3,4)=16.*X(1)*X(3)*(X(2)**2)*X(4)+2.*X(5)*X(2)-2.
      A(3,5)=2.*X(2)*X(4)
      A(4,4)=2.*X(1)*((4.*X(3)2**-2.)*(X(2)**2)+6.*X(4)**2)+2.*X(5)
```

Table 6-1—*Cont.*

```
A(4,5)=2.*X(4)+2.*X(3)*X(2)
RETURN
END
```

The NLP problem was solved for different values of the weighting factor c.

The SUMT program computations were performed on the IBM 360/91 Computer. The results obtained are tabulated as follows:

c	.01	.02	.05	.10	.15	.18	.20	.50	1.00
x_{min}	4.00	4.00	.910	.312	.128	.081	.066	.064	.064
K_{max}	.25	.25	1.100	3.200	7.810	12.35	15.15	15.61	15.61
δ	.5	.5	.612	.654	.676	.684	.687	.687	.687
ω_n	.5	.5	1.414	1.861	2.420	2.783	2.972	2.999	3.000
q	3.0	3.0	1.46×10^{-4}	4.92×10^{-5}	1.82×10^{-5}	1.50×10^{-5}	1.95×10^{-5}	6.40×10^{-5}	2.49×10^{-5}

$a = 1$ for all cases

For each value of c, the run time for the solution of the NLP problem was about 1.30 sec.

The SUMT computation of one of the examples, namely, the one for which $c = .01$, will now be presented in more detail. The results are summarized as follows (see Sec. 3.3):

i	r_i	$f(\mathbf{x})$	$x = x(1)$	$a = x(2)$	$\delta = x(3)$	$\omega_n = x(4)$	$q = x(5)$
	Initial point	.601	.100	.200	.600	.100	.100
0	1.0000	3.736	30.406	.859	.695	2.273	4.901
1	.2500	1.483	78.791	.865	.695	2.256	4.596
2	.0625	.913	21.817	.881	.695	2.186	3.402
3	.0156	.607	8.197	.939	.525	.537	4.223
4	.0039	.556	5.057	.975	.505	.497	3.375
5	9.8×10^{-4}	.544	4.262	.992	.501	.498	3.096
6	2.4×10^{-4}	.541	4.066	.998	.500	.499	3.024
7	6.1×10^{-5}	.540	4.016	.999	.500	.500	3.006
8	1.5×10^{-5}	.540	4.004	1.000	.500	.500	3.001
9	3.8×10^{-6}	.540	4.002	1.000	.500	.500	3.001

As we can see, the program converged after nine iterations. A close approach to the optimal value of the objective function was obtained at the fifth iteration.

EXAMPLE 5

A Fourth-Order System [17]

The general form of the fourth-order system was chosen to be

$$G(s, \mathbf{p}) = \frac{1}{(s^2 + 2\delta_1 \omega_{n1} s + \omega_{n1}^2)(s^2 + 2\delta_2 \omega_{n2} s + \omega_{n2}^2)} \tag{6-38}$$

$$G_1(\omega, \mathbf{p}) = \frac{1}{D}[(\omega_{n1}^2 - \omega^2)(\omega_{n2}^2 - \omega^2) - 4\delta_1 \delta_2 \omega_{n1} \omega_{n2} \omega^2] \tag{6-39}$$

$$G_2(\omega, \mathbf{p}) = -\frac{1}{D}[2\delta_2 \omega_{n2} \omega(\omega_{n1}^2 - \omega^2) + 2\delta_1 \omega_{n1} \omega(\omega_{n2}^2 - \omega^2)] \tag{6-40}$$

where

$$D = (\omega_{n1}^2 - \omega^2)^2(\omega_{n2}^2 - \omega^2)^2 + 4\delta_2^2 \omega_{n2}^2 \omega^2 (\omega_{n1}^2 - \omega^2)^2$$
$$+ 4\delta_1^2 \omega_{n1}^2 \omega^2 (\omega_{n2}^2 - \omega^2)^2 + 16\delta_1^2 \delta_2^2 \omega_{n1}^2 \omega_{n2}^2 \omega^4$$

The adjustable parameter vector in this case is

$$\mathbf{p} = \begin{bmatrix} \delta_1 \\ \omega_{n1} \\ \delta_2 \\ \omega_{n2} \end{bmatrix}$$

Substituting Eqs. (6-39) and (6-40) into Eq. (6-38), in the same manner as for the third-order system, we obtain the following stability inequality constraints:

$$(2\delta_1^2 - 1)\omega_{n1}^2 + (2\delta_2^2 - 1)\omega_{n2}^2 \geqslant 0 \tag{6-41}$$

$$x[4\omega_{n1}^2 \omega_{n2}^2 (4\delta_1^2 \delta_2^2 - 2\delta_1^2 - 2\delta_2^2 + 1) + \omega_{n1}^4 + \omega_{n2}^4]$$
$$+ 1 - 2q(\delta_1 \omega_{n1} + \delta_2 \omega_{n2}) \geqslant 0 \tag{6-42}$$

$$2x\omega_{n1}^2 \omega_{n2}^2[(2\delta_1^2 - 1)\omega_{n2}^2 + (2\delta_2^2 - 1)\omega_{n1}^2] + 2\omega_{n1} \omega_{n2} q(\delta_1 \omega_{n2} + \delta_2 \omega_{n1})$$
$$- \omega_{n1}^2 - \omega_{n2}^2 - 4\delta_1 \delta_2 \omega_{n1} \omega_{n2} \geqslant 0 \tag{6-43}$$

The performance index was

$$\text{Minimize } J = cx + \delta_1 + \delta_2 \tag{6-44}$$

The additional constraints were

$$.5 \leqslant \delta_1 \leqslant 2$$
$$.5 \leqslant \delta_2 \leqslant 2$$
$$\omega_{n1} \leqslant 3 \text{ rad/sec}$$
$$\omega_{n2} \leqslant 3 \text{ rad/sec}$$

The results were as follows:

c	.01	.05	.10	.20	1.0
x_{min}	.251	.163	.138	.121	.100
K_{max}	3.98	6.14	7.25	8.26	10.0
δ_1	.708	.710	.712	.714	.723
ω_{n1}	2.999	2.999	3.0	3.0	3.0
δ_2	.5	.5	.5	.5	.5
ω_{n2}	.236	.398	.496	.611	.916
q	4.748	3.043	2.562	2.204	1.728

For each value of c, the run time for the solution of the NLP problem was about 1.45 sec. Each problem had six variables (x, δ_1, ω_{n1}, δ_2, ω_{n2}, q) and nine inequality constraints.

The algorithm presented above, which consists of a combination of Popov's stability criterion for nonlinear systems and of a NLP optimization technique, may be effectively applied to the control of third- and fourth-order nonlinear systems working under varying performance specifications. It is easy to see that the same algorithm may easily be extended to higher-order systems. There is no restriction whatsoever on the performance index itself—any performance index may be used.

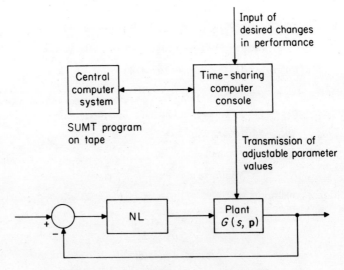

FIG. 6-3. Computer control scheme [17].

With the time-sharing equipment presently available, the actual control could be accomplished in the following way. The SUMT program, with the additional subroutines simulating the system constraints, can be put on tape. Using a remote operation console one can at any time transmit the input data, including the particular changes in the performance index. The actual run takes only a few seconds.

The computed adjustable parameters can then be transmitted directly into the system from the console, in the case of direct digital control, or adjusted by the operator (Fig. 6-3).

6.5 THE INTERIOR PENALTY METHOD

An interior penalty method for the solution of constrained optimal-control problems was presented by Lasdon, Waren, and Rice [19]. The main idea of the method is based on the SUMT algorithm of Fiacco and McCormick (see Chapter 3).

The problem treated by Lasdon et al., is the following:
Given is a system governed by the following set of state equations:

$$\dot{\mathbf{y}} = f(\mathbf{y}, \mathbf{u}, t) \tag{6-45}$$

They are the same as in Eq. (4-1). The initial state vector $\mathbf{y}(t_0) = \mathbf{y}_0$ is assumed known. The terminal region is defined by a set of equality constraints:

$$\mathbf{h}(\mathbf{y}_f, t_f) = \mathbf{0} \tag{6-46}$$

where $\mathbf{h}$ is an r-dimensional vector function. At all times the system should satisfy a set of inequality constraints:

$$\mathbf{g}(\mathbf{y}, \mathbf{u}, t) \geqslant \mathbf{0} \tag{6-47}$$

where $\mathbf{g}$ is an s-dimensional vector function.

The performance index considered is of the terminal control type since it depends explicitly on the final state and time only:

$$\text{Minimize } J = H(\mathbf{y}_f, t_f) \tag{6-48}$$

To solve the problem, the following modified performance index is formulated:

$$\text{Minimize } P(\mathbf{u}, r) = H(\mathbf{y}_f, t_f) + r \sum_{i=1}^{s} \int_{t_0}^{t_f} \frac{dt}{g_i(\mathbf{y}, \mathbf{u}, t)} \tag{6-49}$$

This expression is similar in nature to the one in Eq. (3-129). As in the SUMT algorithm, one would solve a sequence of minimizing $P(\mathbf{u}, r_i)$ going through decreasing values of the r_i coefficients. Between each successive value of r_i, the unconstrained minimization method of conjugate gradients was used [20]. A similar conjugate-gradient solution for optimal-control problems has also been worked out by Sinnott and Luenberger [21].

6.6 SOLUTION OF TWO-POINT BOUNDARY-VALUE PROBLEMS

It is well known that the maximum-principle technique (reference 12 in Chapter 4) involves solving a two-point boundary-value problem (references 1–3 in Chapter 4). The problem is particularly difficult when the maximum principle is applied to the optimal control of nonlinear systems. Rosen and Meyer [22, 23] have proposed applying linear programming directly in the solution of TPBVP.

A typical TPBVP resulting from applying the maximum principle to the solution of optimal-control problems can be formulated as follows:

Given is a set of n ordinary nonlinear first-order differential equations

$$\dot{y}_i(t) = f_i[\mathbf{y}(t), t], \qquad i = 1, \ldots, n \tag{6-50}$$

in the time interval

$$t_0 \leqslant t \leqslant t_f,$$

where the $y_i(t)$'s are the desired functions and the f_i's are known nonlinear continuous functions of $\mathbf{y}(t)$ and t:

$$\mathbf{y}(t) = [y_1(t), \ldots, y_n(t)]^T$$

The boundary conditons are specified as follows:

$$y_i(t_0) = b_{i0}, \qquad i = 1, \ldots, m; m < n \tag{6-51}$$

$$y_i(t_f) = b_{if}, \qquad i = m+1, \ldots, n \tag{6-52}$$

At this point, the Eq. (6-50) can be discretized as described in Chapter 4 or in the previous sections of this chapter. The discretization scheme used by Rosen and Meyer [22, 23] is described as follows using Eq. (6-50):

$$\mathbf{y}(j+1) - \mathbf{y}(j) = \tfrac{1}{2}T\{\mathbf{f}[\mathbf{y}(j+1), j+1] + \mathbf{f}[\mathbf{y}(j), j]\}, \\ j = 0, 1, \ldots, N-1 \tag{6-53}$$

where

$$T = \frac{t_f - t_0}{N}$$

$N =$ number of discrete time subintervals within $t_0 \leq t \leq t_f$
(the subintervals are assumed equal)

$$\mathbf{y}(j) = \mathbf{y}(t_j), t_j = t_0 + jT$$

and

$$f[y(j), j] = f[y(t_j), t_j]$$

We now have a set of nN nonlinear equations with the following unknown variables:

$y_1(1), y_2(2), \ldots, y_i(N), i = 1, \ldots, m$: mN variables

$y_i(0), y_i(1), \ldots, y_i(n-1), i = m+1, \ldots, n$: $(n-m)N$ variables

a total of nN variables

This problem can be represented and solved as a NLP problem by formulating a suitable performance index. For instance, the Hamiltonian can be formulated as a function of y_i appearing in Eq. (6-50) and in turn expressed in discretized form and maximized at each time instant considered. Rosen and Meyer [22, 23] simplified the problem further by linearizing Eq. (6-53); thus LP can be applied.

REFERENCES

1. D. Tabak, "A Direct and Nonlinear Programming Approach to the Optimal Nuclear Reactor Shutdown Control," *IEEE Trans. Nucl. Sci.*, **NS-15**, pp. 57–59, 1968.

2. D. Tabak, "Application of Mathematical Programming in the Design of Optimal Control Systems," Ph.D. Thesis, University of Illinois, Urbana, 1967.

3. M. Ash, *Optimal Shutdown Control of Nuclear Reactors*, Academic Press, New York, 1966.

4. J. J. Roberts and H. P. Smith, "Time Optimal Solution to the Reactivity-Xenon Shutdown Problem," *Nucl. Sci. Eng.*, **22**, pp. 470–478, 1965.

5. J. J. Roberts and H. P. Smith, "Equivalence of the Time Optimal and Minimax Solutions to the Xenon Shutdown Problem," *Nucl. Sci. Eng.*, **23**, pp. 397–399, 1965.

6. Z. R. Rosztoczy and L. E. Weaver, "Optimum Reactor Shutdown Program for Minimum Xenon Buildup," *Nucl. Sci. Eng.*, **20**, pp. 318–323, 1964.

7. R. Bellman, *Dynamic Programming*, Princeton University Press, Princeton, N.J., 1957.

8. B. L. Pierson, "A Discrete Variable Approximation to Optimal Flight Paths," *Astronautica Acta*, **14**, pp. 157–169, 1969.

9. B. L. Pierson, "Numerical Solution of Nonlinear Optimal Control Problems by Discrete Variable Approximation," SIAM 1968 Fall Meeting, Philadelphia, Pa., Oct. 1968.

10. L. Collatz, *The Numerical Treatment of Differential Equations*, 3rd ed., Springer, Berlin, 1960, p. 85.

11. R. Courant, "Variational Methods for the Solution of Problems of Equilibrium and Vibrations," *Bull. Am. Math. Soc.*, **49**, pp. 1–23, 1943.
12. J. Kowalik and M. R. Osborne, *Methods for Unconstrained Optimization Problems*," American Elsevier, New York, 1968.
13. R. Fletcher and C. M. Reeves, "Function Minimization by Conjugate Gradients," *Computer J.*, **7**, pp. 149–154, 1964.
14. D. J. Wilde and C. S. Beightler, *Foundations of Optimization*, Prentice-Hall, Englewood Cliffs, N.J., 1967.
15. D. J. Wilde, *Optimum Seeking Methods*, Prentice-Hall, Englewood Cliffs, N.J., 1964.
16. W. C. Davidon, "Variable Metric Method for Minimization," Report, ANL-5990, Argonne National Laboratory, Argonne, Ill., 1959.
17. D. Tabak, "Computer Control of Nonlinear Systems with Varying Performance Criteria," "*Intern. J. Control*, **11**, pp. 941–947, 1970.
18. V. M. Popov, "Absolute Stability of Nonlinear Systems of Automatic Control," *Automation Remote Control*, **22**, pp. 857–875, Aug. 1961.
19. L. S. Lasdon, A. D. Waren, and R. K. Rice, "An Interior Penalty Method for Inequality Constrained Optimal Control Problems," *IEEE Trans. Automatic Control*, **AC-12**, pp. 388–395, 1967.
20. L. S. Lasdon, S. K. Mitter, and A. D. Waren, "The Conjugate Gradient Method for Optimal Control Problems," *IEEE Trans. Automatic Control*, **AC-12**, pp. 132–138, 1967.
21. J. F. Sinnott, Jr., and D. G. Luenberger, "Solution of Optimal Control Problems by the Method of Conjugate Gradients," 1967 JACC, pp. 566–574, Philadelphia, Pa.
22. J. B. Rosen and R. Meyer, "Solution of Nonlinear Two-Point Boundary Value Problems by Linear Programming," *Computer Sciences Technical Report* #1, University of Wisconsin, Madison, Jan. 1967.
23. J. B. Rosen and R. Meyer, "Solution of Nonlinear Two-Point Boundary Value Problems by Linear Programming," in *Mathematical Theory of Control* (A. V. Balakrishnan and L. W. Neustadt, eds.), Academic Press, New York, 1967, pp. 71–84.

7

LINEAR DISCRETE-TIME SYSTEMS

7.1 INTRODUCTION

A linear discrete-time system with uniform sampling can be described by the following set of state difference equations:

$$\mathbf{y}(k+1) = A\mathbf{y}(k) + B\mathbf{u}(k) \qquad (7\text{-}1)$$

where $\mathbf{y}(k) = n$-dimensional state vector at the discrete time $t = t_k$
$\mathbf{u}(k) = m$-dimensional control vector at $t = t_k$
$A = $ Constant $n \times n$ matrix
$B = $ Constant $n \times m$ matrix

The state vector, at any time $t = t_k$, can be expressed as a function of the initial state and all of the previous control vectors [1]:

$$\mathbf{y}(k) = A^k \mathbf{y}(0) + \sum_{i=0}^{k-1} A^{k-i-1} B\mathbf{u}(i) \qquad (7\text{-}2)$$

Consider now the MP formulation of an optimal-control problem, as given in Eq. (4-26). From Eq. (7-2) we see that the equality constraints of the problem in Eq. (4-26), based on the system's state equations, are now linear. This linearity of the constrained equations is exact, since no approximation is involved. Moreover, the state vector $\mathbf{y}(i)$, at any time $t = t_i$ ($i = 1, \ldots, N$), can be expressed as a function of the initial state vector $\mathbf{y}(0)$, usually specified, and one of the sequence of control vectors $u(i)$ ($i = 0, 1, \ldots, N-1$). This reduces the total number of variables to be considered in the

MP problem by nN, which may considerably improve the efficiency of the numerical solution. If all additional constraints are linear functions of the state and/or control variables, the optimal-control problem involving a system described by Eqs. (7-1) or (7-2) is actually an LP or a QP problem, depending on the formulation of the performance index (see Chapter 2).

Thus it is quite natural that the first optimal-control problems to be treated by MP techniques [2, 3] belong to the class of problems treated in this section. Numerous publications, dealing with applications of MP in optimal control, considered systems of the type described by Eqs. (7-1) or (7-2) [4–14]. A similar class of discrete-time, nonuniformly sampled systems, where the sampling intervals are unknown a priori, was recently solved using LP and QP [15, 16].

The following sections present several specific examples of applying LP and QP methods to the optimal control of linear discrete-time systems.

7.2 EXAMPLES OF MINIMIZATION PROBLEMS

In this section we consider several examples of optimal control of linear discrete-time, uniformly sampled systems. The state equations in all the examples are described by Eq. (7-1), and the expressions used in the equality constraints of the MP formulation are of the form of Eq. (7-2). These various problems differ primarily in their performance indices, and they are classified on this basis.

Minimum-Fuel Problems [2, 3, 6, 8]

The performance index for minimum-fuel problems for discrete-time systems can be expressed as

$$\text{Minimize } J = \sum_{i=0}^{N-1} \sum_{j=1}^{m} c_{ij} |u_j(i)| \qquad (7\text{-}3)$$

where c_{ij} = Constant coefficient
N = Total number of sampling intervals considered
$u_j(i)$ = jth component of the control vector $\mathbf{u}(i)$ at the time $t = t_i$

As discussed in Chapter 5, the performance index of the type in Eq. (7-3) is not a linear expression; therefore, a LP algorithm is not directly applicable. To make it appear to be linear, several methods of modifying this performance index have been described in Chapter 5. Similar methods have also been used in the work devoted primarily to discrete-time systems.

For instance, Whalen [2] considers a case with a scalar control signal $u(i)$ ($i = 0, \ldots, N-1$). His performance index is

$$\text{Minimize } J = \sum_{i=0}^{N-1} c_i |u(i)| \qquad (7\text{-}4)$$

New variables $g(0), \ldots, g(N-1)$ are introduced, so that

$$\left.\begin{array}{r}-g(i) \leqslant u(i) \\ g(i) \geqslant u(i)\end{array}\right\} \quad i = 0, \ldots, N-1 \tag{7-6}$$

Therefore, the new performance index is linear:

$$\text{Minimize } J = \sum_{i=0}^{N-1} c_i g_i \tag{7-6}$$

The reader will recognize this as the bounding variables method discussed in Sec. 5.4.

A different approach was proposed by Torng [6]. He also considered a scalar control signal, and his minimum-fuel performance index is identical to that in Eq. (7-4). The original sequence of the control variable $u(0), u(1), \ldots, u(N-1)$ is replaced by the sequence $u_m(0), u_m(1), \ldots, u_m(N-1)$ so that

$$\left.\begin{array}{r}u_m(i) = u(i) + 1 \\ |u(i)| \leqslant 1\end{array}\right\} \quad i = 0, 1, \ldots, N-1 \tag{7-7}$$

$$0 \leqslant u_m(i) \leqslant 2$$

In addition, the following slack variables $q(i)$ are introduced so that

$$u_m(i) + q(i) = 2, \quad i = 0, 1, \ldots, N-1 \tag{7-8}$$

The minimum-fuel performance index becomes

$$\text{Maximize } J = \sum_{i=0}^{N-1} [c_{ui} u_m(i) + c_{qi} q(i)] \tag{7-9}$$

where

$$c_{ui} = \begin{cases} 1 & \text{if } u_m(i) < 1 \\ 0 & \text{if } u_m(i) = 1 \\ -1 & \text{if } u_m(i) > 1 \end{cases}$$

$$c_{qi} = \begin{cases} 1 & \text{if } q(i) < 1 \\ 0 & \text{if } q(i) = 1 \\ -1 & \text{if } q(i) > 1 \end{cases}$$

Minimum Quadratic Performance Index Problems [10, 11, 14]

A typical form for a quadratic performance index for the system under consideration is [14]

$$\text{Minimize } J = \sum_{i=1}^{N} [\mathbf{y}^T(i) Q \mathbf{y}(i) + \mathbf{u}^T(i-1) R \mathbf{u}(i-1)] \tag{7-10}$$

where Q and R are constant $n \times n$ and $m \times m$ matrices, respectively.

For example, consider the following second-order system [14]:

$$\begin{bmatrix} y_1(k+1) \\ y_2(k+1) \end{bmatrix} = \begin{bmatrix} 1 & 1-e^{-T} \\ 0 & e^{-T} \end{bmatrix} \begin{bmatrix} y_1(k) \\ y_2(k) \end{bmatrix} + \begin{bmatrix} e^{-T}+T-1 & 0 \\ 1-e^{-T} & 0 \end{bmatrix} \begin{bmatrix} u_1(k) \\ u_2(k) \end{bmatrix},$$

$$k = 0, 1, \ldots, 15 \tag{7-11}$$

It is easy to see that Eq. (7-11) takes the same form as Eq. (7-1); in this case, $n = m = 2$. The maximum number of sampling intervals considered is $N = 16$, and the sampling period is $T = .25$ sec. The initial state is given to be

$$\mathbf{y}(0) = \begin{bmatrix} 1 \\ 1 \end{bmatrix} \qquad (7\text{-}12)$$

and the final target state is the origin of the state plane:

$$\mathbf{y}(16) = \begin{bmatrix} 0 \\ 0 \end{bmatrix} \qquad (7\text{-}13)$$

In addition, the control signals are constrained as follows:

$$-1 \leqslant u_j(i) \leqslant 1, \quad j = 1, 2; \, i = 0, 1, \ldots, 15 \qquad (7\text{-}14)$$

The performance index is

$$\text{Minimize } J = \frac{1}{2} \sum_{k=1}^{16} [y_1^2(k) + y_2^2(k) + u_1^2(k-1)] \qquad (7\text{-}15)$$

We see that all the constraints to be derived from Eqs. (7-11) through (7-15) are linear, while the performance index in Eq. (7-15) is quadratic. From this, we know that we have a QP problem. In this case Cutler's RSQPF4 algorithm [17] is used on an IBM 7094 Computer. The results for the control signal $u_1(i)$ are as follows:

i	$u_1(i)$
0	−1.000000
1	−1.000000
2	−1.000000
3	−1.000000
4	−1.000000
5	−.905836
6	−.739215
7	−.608510
8	−.502747
9	−.411670
10	−.324881
11	−.230901
12	−.116078
13	+.036745
14	+.249919
15	+.553175

Figure 7-1 summarizes the results.

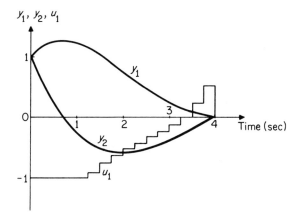

FIG. 7-1. Results of the minimum quadratic performance index problem [14].

Minimum-Time Problems [6, 10, 11]

In a manner similar to the solution of the time-optimal problem in Sec. 5.6, we solve the same problem for the systems of the class considered by a set of sequential subsolutions, utilizing LP or QP techniques. To illustrate this approach, consider Torng's solution [6]. We consider linear, uniformly sampled systems, and at the end of the formulation we obtain a set of linear constraints of the form

$$
\begin{aligned}
a_{11}x_1 + \cdots + a_{1n}x_n + q_1 &= b_1 \\
&\vdots \\
a_{k1}x_1 + \cdots + a_{kn}x_n + q_k &= b_k \\
a_{k+1,1}x_1 + \cdots + a_{k+1,n}x_n &= b_{k+1} \\
&\vdots \\
a_{m1}x_1 + \cdots + a_{mn}x_n &= b_m
\end{aligned}
\qquad (7\text{-}16)
$$

See Sec. 3.1.

The variables $q_1, \ldots, q_k$ in Eq. (7-16) are the *slack variables* (see Sec. 3.1) since the first k equations of Eq. (7-16) are inequalities initially, while the last $m - k$ equations are equalities from the start. To obtain an $m \times m$ unity matrix as a basic feasible solution, we must introduce $m - k$ additional *artificial variables*, so that the last $m - k$ equations of Eq. (7-16) become

$$a_{k+1,1}x_1 + \cdots + a_{k+1,n}x_n + a_1 = b_{k+1}$$
$$\vdots \qquad \vdots \qquad \vdots \qquad \vdots \qquad (7\text{-}17)$$
$$a_{m1}x_1 + \cdots + a_{mn}x_n + a_{m-k} = b_m$$

The first k equations of Eq. (7-16) remain the same.

Now, the following LP subproblem is formulated:

$$\text{Maximize } J = -\sum_{i=1}^{m-k} a_i \qquad (7\text{-}18)$$

subject to all the specified constraints. Eventually, the artificial variables should not appear in the basis.

We try to solve this problem for $N = 1$. If a feasible solution is found, it means that the minimum time is

$$t_{\min} = T$$

If a feasible solution cannot be obtained while eliminating the artificial variables, it means that the terminal state cannot be reached within $t = T$. We should then try for $N = 2$, and so on, until for some $N = N_0$ we obtain a feasible solution which for the first time does not contain the artificial variables. In this case the minimum time is

$$t_{\min} = N_0 T$$

A direct approach to the solution of time-optimal problems using MP is given in references 15, 18, and 19 (see Chapter 6). However, in that case NLP had to be used.

In the works cited [6, 10, 11] more than one admissible control sequence may exist, thus bringing the system into the desired target set in a minimum number of sampling instants N_0. An auxiliary minimization problem is solved after obtaining N_0. In reference 6, a minimum-fuel problem is solved, while in references 10 and 11, an additional minimum-energy (quadratic performance index) problem is solved, and the optimal-control vector sequence is chosen.

Minimum-Error Problems [2, 4]

Whalen [2] considered a problem of minimizing the error between the final state of the system $\mathbf{y}(N)$ and a prescribed desired state $\mathbf{y}_d$. The error vector $\mathbf{e}$, whose dimension is assumed to be K, is defined as follows:

$$\mathbf{e} = D[\mathbf{y}(N) - \mathbf{y}_d] \qquad (7\text{-}19)$$

where D is a $(K \times n)$-dimensional constant matrix. The performance index is then formulated as follows:

$$\text{Minimize } J = \sum_{i=1}^{K} e_i \qquad (7\text{-}20)$$

Clearly the performance index in Eq. (7-20) is a linear function of the state variables of the system at the final time $t = t_N$. Using Eq. (7-2), we see that the performance index is also a linear function of the sequence of the control vectors $\mathbf{u}(i)$. If all the other constraints of the problem are linear, an LP algorithm is readily applicable.

A different performance index was used by Propoi [4]. One of the state variables, y_1, is identified as the output variable. The system error at any time $t = t_i$ is defined as

$$e_i = y_0(i) - y_1(i) \qquad (7\text{-}21)$$

where $y_0(i)$ is a prescribed reference state, which in general may be time varying.

The performance index is formulated as

$$\text{Minimize } J = \sum_{i=1}^{N} e_i \qquad (7\text{-}22)$$

Again, we obtain a linear function of the state variables and hence of the control variables. To prevent the possibility of overshoot, a special set of constraints is imposed:

$$e_i \geqslant 0, \quad i = 1, \ldots, N \qquad (7\text{-}23)$$

Otherwise, we would have to use a quadratic performance index.

7.3 DESIGN OF A DIGITAL CONTROLLER

This section describes the application of LP and QP in the design of a digital controller for a linear, uniformly sampled-data control system as proposed by Porcelli and Fegley [5, 7, 9].

The basic configuration of the system considered is shown in Fig. 7-2.

The plant is represented by its z-transformed transfer function $G(z)$. Similarly, $D(z)$ represents the digital controller, and $K(z)$ is the overall closed-loop transfer function. The input to the system is specified by the following pulse sequence:

$$R(z) = r_0 + r_1 z^{-1} + r_2 z^{-2} + \cdots + r_n z^{-n} = \sum_{i=0}^{n} r_i z^{-i} \qquad (7\text{-}24)$$

where $r_i = r(t_i)$ denotes the known amplitudes of the input pulse at the sampling instant t_i ($i = 0, 1, \ldots, n$). Similarly, the overall transfer function of the system is represented by

$$K(z) = A_m z^{-m} + A_{m+1} z^{-(m+1)} + \cdots + A_{m+k} z^{-(m+k)} \qquad (7\text{-}25)$$

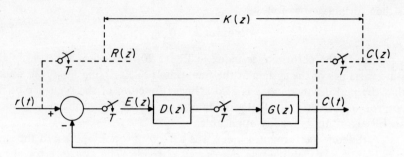

FIG. 7-2. Block diagram of the sampled data system [9].

where m represents the inherent delay in the system. Because the digital controller $D(z)$ is unknown a priori, the coefficients A_i ($i = m, \ldots, m + k$) of the overall transfer function are also unknown.

The basic equations of the system shown in Fig. 7-2 are

$$C(z) = K(z)R(z) \qquad (7\text{-}26)$$

$$E(z) = R(z) - C(z) \qquad (7\text{-}27)$$

$$K(z) = \frac{C(z)}{R(z)} = \frac{D(z)G(z)}{1 + D(z)G(z)} \qquad (7\text{-}28)$$

From Eq. (7-28) we can express $D(z)$ as

$$D(z) = \frac{K(z)}{G(z)[1 - K(z)]} \qquad (7\text{-}29)$$

The plant $G(z)$ is usually specified a priori. Once $K(z)$ is established, we can find the expression for the digital controller $D(z)$ by using Eq. (7-29). In the following, the method of establishing the A_i coefficients of $K(z)$ using LP or QP is discussed.

Combining Eqs. (7-26) and (7-27) and using Eqs. (7-24) and (7-25) in the resulting equation, we can then equate the coefficients of equal powers in z^{-1} to produce the following matrix-vector equation:

$$\mathbf{RA} + \mathbf{e} = \mathbf{r} \qquad (7\text{-}30)$$

where

$$\mathbf{e} = \begin{bmatrix} e_m \\ e_{m+1} \\ \cdot \\ \cdot \\ \cdot \\ e_n \end{bmatrix} \qquad \mathbf{r} = \begin{bmatrix} r_m \\ r_{m+1} \\ \cdot \\ \cdot \\ \cdot \\ r_n \end{bmatrix} \qquad \mathbf{A} = \begin{bmatrix} A_m \\ A_{m+1} \\ \cdot \\ \cdot \\ \cdot \\ A_{m+k} \end{bmatrix}$$

Sec. 7.3   Design of a Digital Controller   **143**

$$R = \begin{bmatrix} r_0 & 0 & 0 & \cdots & 0 \\ r_1 & r_0 & 0 & \cdots & 0 \\ r_2 & r_1 & r_0 & \cdots & 0 \\ \vdots & \vdots & \vdots & & \vdots \\ r_k & r_{k-1} & r_{k-2} & \cdots & r_0 \\ \vdots & \vdots & \vdots & & \\ r_{n-m} & r_{n-m-1} & r_{n-m-2} & \cdots & r_{n-m-k} \end{bmatrix} \quad (n+1-m) \times (k+1)$$

The error pulses e_i ($i = m, \ldots, n$) along with the A_i ($i = m, \ldots, m+k$) coefficients are the unknown variables of the problem. We see that Eqs. (7-30) constitute a set of linear equality constraints of $(n + 1 - m)$ equations in $(n + k - m + 2)$ variables. In the established LP algorithms it is customary to assume nonnegativity of all the variables (see Chapter 3). This is not the case here, since e_i and A_i can be positive as well as negative. To make the actual variables of the problem nonnegative, we must introduce the following variables:

$$A_i = A_i^+ - A_i^- \quad \text{for all } i \qquad (7\text{-}31)$$
$$e_i = e_i^+ - e_i^- \quad \text{for all } i \qquad (7\text{-}32)$$

where

$$A_i^+, A_i^-, e_i^+, e_i^- \geqslant 0 \quad \text{for all } i \qquad (7\text{-}33)$$

Substituting Eqs. (7-31) and (7-32) into Eq. (7-30), we obtain

$$RA^+ - RA^- + e^+ - e^- = r \qquad (7\text{-}34)$$

In addition to the constraints in Eqs. (7-32) and (7-34), the following constraints are posed:

$$-E_m \leqslant (e_i^+ - e_i^-) \leqslant E_m \quad \text{for all } i \qquad (7\text{-}35)$$

To implement an LP algorithm, a linear objective function must be introduced. Initially, the objective function was formulated as the weighted sum of the absolute values of the errors:

$$\text{Minimize } J = \sum_{i=m}^{n} c_i |e_i| = \sum_{i=m}^{n} c_i |e_i^+ - e_i^-| \qquad (7\text{-}36)$$

The objective function in Eq. (7-36) is further simplified according to the following considerations:

As mentioned in Chapter 3, the optimal solution of an LP problem of m constraint equations in n unknowns ($m < n$), if any feasible solution exists, is always a basic solution, i.e., a solution with not more than m nonzero values for the variables. This means that the **b** vector in Eq. (3-1) or Eq. (3-3) can be written as a linear combination of m column vectors from the matrix A in Eq. (3-1). These m vectors of m dimensions are called basic vectors and constitute a basis for the m-dimensional space. If the variables of an LP problem are unrestricted in sign, only one of the two components of each variable can appear in the optimal solution. This can readily be proved as follows:

Consider, for example, that x_j in the system of Eq. (3-1) is unrestricted. Consequently, the jth column of A yields

$$\mathbf{a}_j x_j^+ + (-\mathbf{a}_j) x_j^-$$

where $\mathbf{a}_j$ is the vector representing the jth column of matrix A. Since $\mathbf{a}_j$ and $-\mathbf{a}_j$ are *not independent*, no basis can contain both of them. Therefore, no optimal solution can contain both x_j^+ and x_j^- [7].

Following these considerations, either e_j^+ or e_j^- or both are zero in the optimal solution. Hence, $|e_j^+ - e_j^-|$ can be replaced by $e_j^+ + e_j^-$, and the objective function can now be rewritten as

$$\text{Minimize } J = \sum_{i=m}^{n} c_i(e_i^+ + e_i^-) \tag{7-37}$$

Equation (7-37) is a perfectly linear expression in terms of the problem variables e_i^+ and e_i^-, and LP algorithms can be directly applied to the numerical solution of this problem.

Alternatively, a quadratic performance index of minimizing the weighted sum of squared errors can be posed [9]:

$$\text{Minimize } J = \sum_{i=m}^{n} c_i(e_i^+ - e_i^-)^2 \tag{7-38}$$

A QP algorithm, RSQPF4 [17], is applied to the numerical solution of this problem. Three examples illustrating the implementation of the method are presented in the following [9].

EXAMPLE 1

A digital-control system of the form of Fig. 7-2 has the plant pulse transfer function

$$G(z) = \frac{(1 + 2.34z^{-1})(1 + .16z^{-1})z^{-1}}{(1 - z^{-1})(1 - .368z^{-1})^2}$$

The system is required to meet the following specifications: (1) The input is a unit ramp function; (2) all variables are unrestricted in sign; (3) the design is based on the first 15 sampling instants, and the sampling period is 1 sec; (4) $K(z)$ is limited to five terms; i.e., $K(z) = A_1 z^{-1} + A_2 z^{-2} + A_3 z^{-3} + A_4 z^{-4} + A_5 z^{-5}$; (5) the constraints are

$$|e_2| \leqslant 1, \quad |e_j| \leqslant .5, \quad j = 3, 4, \ldots, 15$$

It is shown in reference 1 that for a system of the type of Fig. 7-2, with $G(z)$ given, it is necessary that $K(z)$ contain as its zeros all the zeros of $G(z)$ which lie outside or on the unit circle of the z plane, and that $1 - K(z)$ contain as its zeros

Sec. 7.3 Design of a Digital Controller **145**

all the poles of $G(z)$ which lie outside or on the unit circle of the z plane. As shown in reference 7, the constraint equations that follow these two necessary conditions applied to this example are

$$HA^+ - HA^- = \delta$$

where

$$H = \begin{bmatrix} 4.16 & -1.34 & 1 & 0 & .4275 \\ 1 & 1 & 1 & 1 & 1 \end{bmatrix}$$

and

$$\delta = [.3 \quad 1]$$

The problem is solved by first using a purely linear objective function of the form

$$\min J = \min \sum_{i=1}^{15} i(e_i^+ + e_i^-)$$

and then the purely quadratic objective function

$$\min J = \min \sum_{i=1}^{15} (e_i^+ - e_i^-)^2$$

The results for the linear objective function are

$$
\begin{aligned}
A_1 &= 1.000000 & e_1 &= 1.000000 \\
A_2 &= 1.455090 & e_2 &= 1.000000 \\
A_3 &= -1.910180 & e_3 &= -.455090 \\
A_4 &= .455090 &
\end{aligned}
$$

All the other variables are zero. The objective function has the value

$$J = 4.365279$$

The overall pulse transfer function of the system is

$$K(z) = z^{-1} + 1.455090z^{-2} - 1.910180z^{-3} + .455090z^{-4}$$

The results for the quadratic objective function are

$$
\begin{aligned}
A_1 &= 1.018104 & e_6 &= -.016549 \\
A_2 &= 1.382101 & e_7 &= -.013629 \\
A_3 &= -1.997295 & e_8 &= -.010708 \\
A_4 &= .795343 & e_9 &= -.007788 \\
A_5 &= -.201173 & e_{10} &= -.004868 \\
e_1 &= 1.000000 & e_{11} &= -.001948 \\
e_2 &= .981896 & e_{12} &= .000973 \\
e_3 &= -.418308 & e_{13} &= .003893 \\
e_4 &= .178783 & e_{14} &= .006813 \\
e_5 &= .019469 & e_{15} &= .009733
\end{aligned}
$$

The objective function $J = 2.172283$.

The overall pulse transfer function is

$$K(z) = 1.018104z^{-1} + 1.382101z^{-2} - 1.997295z^{-3} + .795343z^{-4} - .201173z^{-5}$$

Results for the quadratic objective function are quite similar to the results for the linear objective function. Comparing them (Fig. 7-3) shows, however, that for this example a purely linear objective function yields a slightly faster response than a purely quadratic objective function.

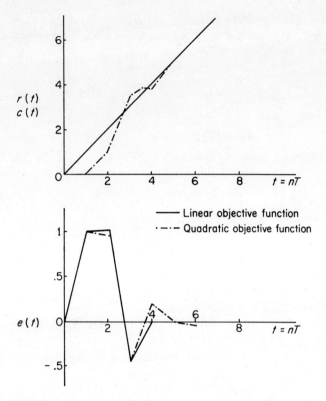

FIG. 7-3. Input, sampled output and sampled error for Example 1 [9].

EXAMPLE 2

A system where the plant does not impose any restrictions on the overall pulse transfer function is designed to meet the following specifications:
1. The input represented in Fig. 7-4 is given by

$$r(t) = .1t^2, \qquad 0 \leqslant t < 5T$$
$$r(t) = .1[50 - (t - 10)^2], \qquad 5T \leqslant t < 15T$$
$$r(t) = .1(t - 20)^2, \qquad 15T \leqslant t < 20T$$
$$r(t) = 0, \qquad 20T \leqslant t$$

2. All variables are unrestricted.
3. The design is based on the first 24 sampling instants.
4. $K(z)$ is limited to three terms and has the form

$$K(z) = A_1 z^{-1} + A_2 z^{-2} + A_3 z^{-3}$$

Two different solutions are given, the first obtained with a purely linear objective function of the form

$$\min J = \min \sum_{i=1}^{24} (e_i^+ + e_i^-)$$

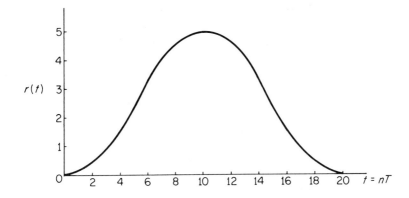

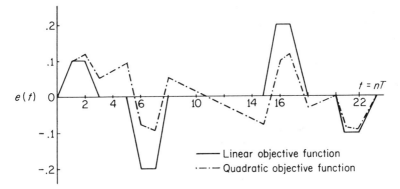

FIG. 7-4. Input and sampled error for Example 2 [9].

and the second with a purely quadratic objective function of the form

$$\min J = \min \sum_{i=1}^{24} (e_i^+ - e_i^-)^2$$

The results for the linear objective function are

$A_1 = 2.999990 \qquad e_1 = .100000$
$A_2 = -2.999982 \qquad e_2 = .100000$
$A_3 = .999992 \qquad e_6 = -.199995$
$ \qquad e_7 = -.199997$
$ \qquad e_{16} = .199997$
$ \qquad e_{17} = .199998$
$ \qquad e_{21} = -.099999$
$ \qquad e_{22} = -.099999$

All other errors are zero.
The objective function $J = 1.200003$.
The overall pulse transfer function is

$$K(z) = 2.999990z^{-1} - 2.999982z^{-2} + .999992z^{-3}$$

148   Linear Discrete-Time Systems   Ch. 7

The results for the quadratic objective function are

$$
\begin{aligned}
A_1 &= 2.763373 & e_{10} &= .024329 \\
A_2 &= -2.632494 & e_{11} &= .005469 \\
A_3 &= .864540 & e_{12} &= -.014306 \\
e_1 &= .100000 & e_{13} &= -.034998 \\
e_2 &= .123663 & e_{14} &= -.056605 \\
e_3 &= .057900 & e_{15} &= -.079129 \\
e_4 &= .079508 & e_{16} &= .097431 \\
e_5 &= .102032 & e_{17} &= .120400 \\
e_6 &= -.074529 & e_{18} &= -.036397 \\
e_7 &= -.097498 & e_{19} &= -.019369 \\
e_8 &= .059299 & e_{20} &= -.001426 \\
e_9 &= .042272 & e_{21} &= -.082567 \\
 & & e_{22} &= -.086454 \\
 & & e_{23} &= e_{24} = 0
\end{aligned}
$$

The objective function $J = 0.117240$.

The overall pulse transfer function is

$$K(z) = 2.763373 z^{-1} - 2.632494 z^{-2} + .864540 z^{-3}$$

Both results are illustrated in Fig. 7-4.

EXAMPLE 3

This example is similar to Example 2; the only difference is that now $K(z)$ has five terms instead of three, and the linear and quadratic objective functions are time weighted. That is,

$$K(z) = A_1 z^{-1} + A_2 z^{-2} + A_3 z^{-3} + A_4 z^{-4} + A_5 z^{-5}$$

$$\min J = \min \sum_{i=1}^{24} i(e_i^+ + e_i^-)$$

$$\min J = \min \sum_{i=1}^{24} i(e_i^+ - e_i^-)^2$$

For the linear objective function, the results are

$$
\begin{aligned}
A_1 &= 2.872985 & e_{13} &= -.025045 \\
A_2 &= -3.457953 & e_{14} &= -.052952 \\
A_3 &= 2.440063 & e_{15} &= -.083721 \\
A_4 &= -1.159208 & e_{16} &= .082647 \\
A_5 &= .289802 & e_{17} &= .071556 \\
e_1 &= .100000 & e_{19} &= .071556 \\
e_2 &= .112702 & e_{20} &= .030054 \\
e_3 &= .096601 & e_{21} &= -.050626 \\
e_4 &= .153489 & e_{22} &= -.041145 \\
e_5 &= .155278 & e_{24} &= -.028980 \\
e_6 &= -.011091 & \multicolumn{3}{l}{e_7 = e_9 = e_{12} = e_{18} = e_{23} = 0} \\
e_8 &= .071556 \\
e_{10} &= .041503 \\
e_{11} &= .022182
\end{aligned}
$$

Sec. 7.3 *Design of a Digital Controller* **149**

The objective function $J = 12.789630$.

$K(z) = 2.872985z^{-1} - 3.457953z^{-2} + 2.440063z^{-3} - 1.159208z^{-4} + .289802z^{-5}$

For the quadratic objective function, the results are

$A_1 = 3.067555$ $e_{10} = .022914$
$A_2 = -3.859141$ $e_{11} = .006429$
$A_3 = 2.623430$ $e_{12} = -.011841$
$A_4 = -1.062250$ $e_{13} = -.031897$
$A_5 = .221480$ $e_{14} = -.053738$
$e_1 = .100000$ $e_{15} = -.077364$
$e_2 = .093244$ $e_{16} = .097225$
$e_3 = .058892$ $e_{17} = .056519$
$e_4 = .120514$ $e_{18} = -.041168$
$e_5 = .121991$ $e_{19} = .051309$
$e_6 = -.052598$ $e_{20} = .021713$
$e_7 = -.011891$ $e_{21} = -.061801$
$e_8 = .085795$ $e_{22} = -.036775$
$e_9 = -.006682$ $e_{23} = .017633$
$e_{24} = -.022148$

The objective function $J = 0.822381$.

$K(z) = 3.067555z^{-1} - 3.859141z^{-2} + 2.623430z^{-3} - 1.062250z^{-4} + .221480z^{-5}$

The sampled errors are shown in Fig. 7-5 for both solutions.

Comparison of Figs. 7-4 and 7-5 shows that the increased number of terms in $K(z)$ yields a smoother error function and that the time weighting in the objective functions yields a faster decreasing error amplitude. The results of these two examples show that the difference between the solutions by LP and by QP decreases as the number of allowed coefficients of the overall pulse transfer function $K(z)$ increases.

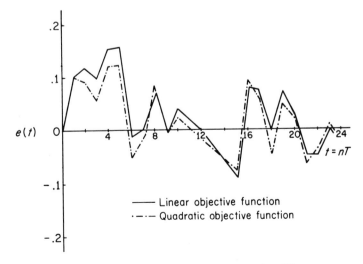

FIG. 7-5. Sampled error for Example 3 [9].

7.4 DISCRETE-TIME-CONTROLLED SYSTEMS WITH UNKNOWN SAMPLING PERIODS

In this section we discuss a linear system which is unforced except at a finite number of time instants [15, 16]. This means that the system may generally be considered as a continuous time system governed by a set of linear state equations of the form

$$\dot{\mathbf{x}} = A\mathbf{x} \qquad (7\text{-}39)$$

where $\mathbf{x} = n$-dimensional state vector
$A = n \times n$ constant matrix

At certain discrete times, $t_1, \ldots, t_N$, a discontinuous change in the values of the state variables occurs. This change is regarded as a control variable, and all the changes for all components of $\mathbf{x}$ form a control vector applied at discrete times $t_1, \ldots t_N$, which are unknown a priori.

The time elapsed between the occurrence of subsequent discontinuities in the state vector or applications of the control are called the continuous period or cycle. Each period has a certain length T_i ($i = 1, \ldots, N$), which is unknown a priori. The time instants $t_1, \ldots, t_N$ at which the control is applied are called sampling (or control) instants. Obviously,

$$T_i = t_i - t_{i-1}, \quad i = 1, \ldots, N \qquad (7\text{-}40)$$

where N is the maximum number of continuous periods or cycles considered.

The state vector at the sampling time $t = t_i$ is designated as $\mathbf{x}(i)$, and the control vector is designated as $\mathbf{u}(i)$. The jth component of the state and control vectors at $t = t_i$ are designated as $x_j(i)$ and $u_j(i)$, respectively.

The time pattern of the state and control vectors along the time axis can be presented schematically in the following way:

Cycles	T_1	T_2			T_{N-1}	T_N
Time	t_0	t_1	t_2	$\cdots$	t_{N-2}	t_{N-1}	t_N
State vector	$\mathbf{x}(0)$	$\mathbf{x}(1)$	$\mathbf{x}(2)$	$\cdots$	$\mathbf{x}(N-2)$	$\mathbf{x}(N-1)$	$\mathbf{x}(N)$
Control vector	$\mathbf{u}(0)$	$\mathbf{u}(1)$	$\mathbf{u}(2)$	$\cdots$	$\mathbf{u}(N-2)$	$\mathbf{u}(N-1)$	$\mathbf{u}(N)$

Because $t = t_N$ is the final time considered which is free, and no further change in the final state $\mathbf{x}(N)$ is considered, $\mathbf{u}(N) = 0$. The initial state $\mathbf{x}(0)$ is assumed known. However, because of reasons which are clarified in further development, it is assumed that $\mathbf{u}(0) = \mathbf{x}(0)$, contrary to the general definition of the control vector as a discrete change of the state vector. This definition will then hold, starting only with $t = t_1$.

The solution of Eq. (7-39) at any time t, starting with $t = t_0$, is

$$\mathbf{x}(t) = e^{A(t-t_0)}\mathbf{x}(t_0) \qquad (7\text{-}41)$$

where $e^{A(t-t_0)}$ is an $n \times n$ matrix.

At the end of the first cycle at $t = t_1$, the state vector x attains a final value:

$$\mathbf{x}_f(1) = \mathbf{x}(t_1) = e^{A(t_1-t_0)}\mathbf{x}(0) = e^{AT_1}\mathbf{x}(0) \qquad (7\text{-}42)$$

At this time the control $\mathbf{u}(1)$ is applied in the following way to obtain the actual value of x at $t = t_1$:

$$\mathbf{x}(1) = \mathbf{x}_f(1) + \mathbf{u}(1) = e^{AT_1}\mathbf{x}(0) + \mathbf{u}(1) \qquad (7\text{-}43)$$

We see that $\mathbf{u}(1)$ introduces a discrete change in the value $\mathbf{x}_f(1)$ of the state vector, and we can express this discontinuity in the following form:

$$\left.\begin{array}{l}\mathbf{x}(t_1 - \epsilon) = \mathbf{x}_f(1) \\ \mathbf{x}(t_1 + \epsilon) = \mathbf{x}(1)\end{array}\right\} \quad \epsilon > 0;\ \epsilon \ll T_1$$

A similar process occurs at all sampling instants so that we can write a general difference equation:

$$\mathbf{x}(i) = e^{AT_i}\mathbf{x}(i-1) + \mathbf{u}(i), \qquad i = 1, \ldots, N \qquad (7\text{-}44)$$

The state difference equation (7-44) depends not only on the control and state vectors, but also on the length of the cycles between the sampling instants. Considering that all the periods T_i ($i = 1, \ldots, N$) are unknown a priori, we face a difficult problem of finding those periods. We propose an iterative algorithm to establish the unknown periods by using other properties and constraints of the system; this algorithm will be described later in this section, and additional properties of the system will be discussed.

As in any optimal-control problem, a performance index will be specified. In this problem we want to minimize or maximize a performance index of the general form

$$J = \sum_{i=1}^{N} F[\mathbf{x}(i), \mathbf{u}(i)] \qquad (7\text{-}45)$$

The continuous periods are not initiated and terminated deliberately; certain constraints dependent on the state and control variables must be satisfied at all times. In particular, at the beginning of each cycle, an initialization constraint must be satisfied:

$$g_{\text{in}}[\mathbf{x}(i), \mathbf{u}(i)] \geqslant 0, \qquad i = 0, \ldots, N-1 \qquad (7\text{-}46)$$

No cycle can be started unless inequality equation (7-46) is satisfied. Also, no cycle in progress can continue unless the following termination constraints are satisfied:

$$g_t[\mathbf{x}_f(i)] \geqslant 0, \qquad i = 1, \ldots, N \qquad (7\text{-}47)$$

Another condition imposed on the system is that it should be operated during the continuous cycles for the longest time possible without violating the constraint, Eq. (7-47). Therefore, at the end of each period, the constraint actually takes the form of an equality. As soon as the condition in Eq. (7-47) is violated, the cycle is terminated, and the control is applied. If the condition in Eq. (7-46) is satisfied after applying the control, a new cycle can be started.

After completing the definition of the problem, the computational algorithm can now be formulated:

1. Use the specified initial condition x(0) and solve Eq. (7-39) numerically. At each integration step, check whether or not $g_i(x) \geqslant 0$. As soon as $g_i(x) < 0$, stop the integration process and record the final time T_1, which is the first sampling period.
2. Set all the sampling time periods equal to T_1:
$$T_1 = T_2^{(1)} = T_3^{(1)} = \cdots = T_N^{(1)}$$
The superscript (1) refers to the first iteration.
3. Substitute $T_i^{(1)}(i = 1, \ldots, N)$ into the difference state equations (7-44). Using Eq. (7-44), we can express every state vector at any sampling instant t_i as a function of the previous and present control vectors $\mathbf{u}(j)$ ($j = 1, \ldots, i$) and the initial vector $\mathbf{x}(0) = \mathbf{u}(0)$ in the following way:
$$\mathbf{x}(i) = \sum_{m=0}^{i-1} \left[\prod_{j=m+1}^{i} \exp(AT_j) \right] \mathbf{u}(m) + \mathbf{u}(i) \qquad (7\text{-}48)$$
Equation (7-48) is then substituted into the performance index in Eq. (7-45) as well as into the constraints in Eqs. (7-46) and (7-47) and any additional constraints.
4. A MP problem, extremizing Eq. (7-45) subject to all constraints, is solved, so that we obtain a set of optimal $\mathbf{u}^{(1)}(m)$ ($m = 1, \ldots, N-1$). In particular, if all the constraints as well as the performance index are linear, the result would be LP problem.
5. Equation (7-39) is now solved for all N periods. Each period is terminated as soon as $g_i(\mathbf{x}) < 0$. Suppose the final value at the end of period i is $\mathbf{x}_f(i)$. Then the initial condition for the next integration is $\mathbf{x}(i) = \mathbf{x}_f(i) + \mathbf{u}(i)$, etc. After solving for all cycles, a net set of periods $T_1^{(2)}, T_2^{(2)}, \ldots, T_N^{(2)}$ is obtained. Note that $T_1^{(2)} = T_1$, since a constant initial vector $\mathbf{x}(0)$ is assumed. In general, we might vary $\mathbf{x}(0)$ and T_1 as well. The algorithm remains the same.
6. Take the new $T_i^{(2)}$ ($i = 1, \ldots, N$), and repeat step steps 3–5, obtaining at the end $T_i^{(3)}$, etc. The process will continue until
$$|T_i^{(k)} - T_i^{(k+1)}| \leqslant e > 0$$
where e is a specified convergence criterion. $\mathbf{u}^{(k)}(i)$ ($i = 1, \ldots, N-1$) would serve as a final control law.

The convergence of this algorithm has been established by computational experimentation.

EXAMPLE

The technique proposed can be applied to a problem of optimal-fuel recycling in nuclear reactors. For illustrational purposes a simplified point model of a thermal nuclear reactor has been chosen (Fig. 7-6). The state variables are now the nuclide concentrations of the materials considered, and the control variables are the discrete changes of those concentrations at refueling times (Fig. 7-7).

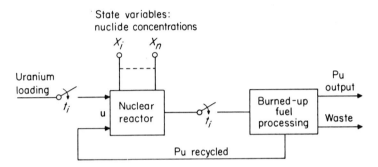

FIG. 7-6. The nuclear reactor fuel cycle model [16].

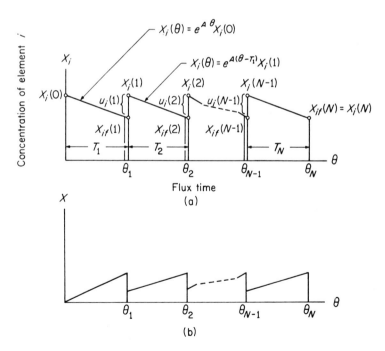

FIG. 7-7. Time change of nuclide i [16]: (a) element undergoing depletion (like U-235); (b) element undergoing generation (like Pu-239).

The state equations, with the flux time as the independent variable, look exactly like Eq. (7-39). Matrix A depends on the nuclear properties of the materials considered, such as the cross sections. A necessary condition for a successful operation of a nuclear reactor is that its multiplication factor $k(\mathbf{x})$ be higher than unity. We usually prescribe a certain minimum value of k to start an irradiation cycle, in this case

$$k_{\text{in}}(\mathbf{x}) \geqslant 1.08 \tag{7-49}$$

in analogy with the constraint, Eq. (7-46). The constraint which determines the termination of each irradiation cycle is

$$k_t(\mathbf{x}) \geqslant 1.0001 \tag{7-50}$$

in analogy with the constraint in Eq. (7-47). As soon as Eq. (7-50) is violated, the reactor must be stopped and refueling must occur.

The basic fuel undergoing fission in the nuclear reactor is U-235, which occurs at .72% in natural uranium. The enriching of the uranium to increase the percentage of U-235 in it is a very expensive process, and we tend to operate the reactor with minimum usage of U-235.

On the other hand, a by-product of the irradiation process in a nuclear reactor is the generation of Pu-239, which can serve as fissionable fuel in the same reactor or which can be extracted and used for other purposes. A problem that arises is to find the optimal combination of U-235 fed in and Pu-239 taken out so that the overall operating cost is reduced.

Two particular performance indices are considered:

P1. Minimization of U-235 usage over the total range of fuel cycles, with a requirement that the overall Pu-239 output not be less than a certain specified amount C. This performance index is expressed both in linear and quadratic form. Linear form:

$$J = \sum_{i=0}^{N-1} [x_1(i) - x_{f1}(i+1)] \tag{7-51}$$

Quadratic form:

$$J = \sum_{i=0}^{N-1} [x_1(i) - x_{f1}(i+1)]^2 \tag{7-52}$$

The subscript 1 refers to U-235, the first component of vector $\mathbf{x}$.

P2. Maximization of Pu-239 output, keeping the U-235 enrichment limited to a prescribed value. The performance index is expressed in linear form:

$$J = \sum_{i=1}^{N-1} u_3(i) + x_{f3}(N) \tag{7-53}$$

The subscript 3 refers to Pu-239, the third component of vector $\mathbf{x}$.

A total of four state variables is considered:

x_1	U-235
x_2	U-238
x_3	Pu-239
x_4	fission fragments

For this model the matrix A of Eq. (7-39) is

$$A = \begin{bmatrix} -\sigma_{a1} & 0 & 0 & 0 \\ -f_1 & -\sigma_{a2} & -f_3 & 0 \\ 0 & \sigma_{a2} & -\sigma_{a3} & 0 \\ \sigma_{f1} & 0 & \sigma_{f3} & -\sigma_{a4} \end{bmatrix}$$

where σ_{ai} = Microscopic absorption cross section of nuclide i
 σ_{fi} = Microscopic fission cross section of nuclide i
 f_i = Fraction of neutrons obtained as a result of the fission of nuclide i which have survived leakage and which are absorbed by U-238

The multiplication factor is linearized to the following form:

$$k = \frac{\sum_{i=1}^{4} \gamma_i \sigma_{f1} x_i}{\sum_{i=1}^{4} \sigma_{ai} x_i + \Sigma_s} \tag{7-54}$$

where γ_i = Number of neutrons produced per fission of nuclide i
 Σ_s = Macroscopic cross section representing all the materials present in the reactor, with the exception of those considered as state variables

The following constraints were considered at each sampling instant $i = 1, \ldots, N - 1$:

C1. Initialization constraint. Using Eqs. (7-49) and (7-54),

$$\sum_{j=1}^{4} \gamma_j \sigma_{fj} x_j(i) - 1.08 \sum_{j=1}^{4} \sigma_{aj} x_j(i) - 1.08 \Sigma_s \geqslant 0 \tag{7-55}$$

C2. Termination constraint. Using Eqs. (7-50) and (7-54),

$$\sum_{j=1}^{4} \gamma_j \sigma_{fj} x_{fj}(i+1) - 1.001 \sum_{j=1}^{4} \sigma_{aj} x_{fj}(i+1) - 1.0001 \Sigma_s \geqslant 0 \tag{7-56}$$

C3. The quantity of Pu-239 to be taken out at the sampling instant i should not exceed the quantity of Pu-239 obtained at the end of the ith cycle:

$$u_3(i) \leqslant x_{f3}(i) \tag{7-57}$$

C4. The uranium enrichment at the beginning of each cycle should not exceed $E_n \%$:

$$\frac{x_1(i)}{x_1(i) + x_2(i)} \leqslant \frac{E_n}{100} \tag{7-58}$$

C5. The overall output of Pu-239 should be at least C (in Prob. Pl only):

$$\sum_{i=1}^{N-1} u_3(i) + x_{f3}(N) \geqslant C \tag{7-59}$$

In all the specified constraints and the performance indices, the state variables are expressed in terms of the control variables $u_j(i)$ using Eqs. (7-43), (7-44), and (7-48). We see that all the constraints have the form of linear inequalities. Therefore, depending on the form of J given in Eqs. (7-51) through (7-53), LP or QP can be used.

Note that since no fission fragments are loaded, $u_4(i) = 0$ for all i; thus, only three control variables are considered at each sampling time. Therefore, for N cycles considered, we have a total of $3(N-1)$ variables with $4(N-1)+1$ constraints for Prob. P1 and $4(N-1)$ for Prob. P2. Since Pu-239 is usually taken out, $u_3(i)$ always appears with a negative sign in the equations loaded into the computer.

Table 7-1 [16]

Nuclide	σ_a (barns)	σ_f (barns)	γ
1. U-235	362.4	309	2.42
2. U-238	1.56	0	0
3. Pu-239	1103.0	683	2.90
4. Fission fragments	2.0	0	0

Note: The values are chosen for illustrational purposes; they do not necessarily represent the actual values existing simultaneously in a practical reactor.

Table 7-2 [16]

Problem no.	1	2	3	4
J	P1, quadratic	P1, linear	P2, linear	P2, linear
$\mathbf{x}(0) =$	3.35 68.11 0 0	3.42 81.90 0 0	3.35 68.11 0 0	3.42 81.90 0 0
$E_i\%$	4.92	4.00	4.92	4.00
N_i	5	1	8	11
$\mathbf{u}_0(1) =$	.64141 14.11885 −.03982 0	.61118 .45596 −.04996 0	.64497 14.20418 −.04197 0	.71842 22.89882 −.05208 0
$\mathbf{u}_0(2) =$	.61668 .42683 −.05321 0	.61355 .29003 −.05088 0	.64131 6.47378 −.05207 0	.74749 17.38654 −.06932 0
$\mathbf{u}_0(3) =$	.61681 .35913 −.05225 0	.60910 .17468 −.04773 0	.64647 4.32040 −.05673 0	.76435 11.63549 −.08330 0
$\mathbf{u}_0(4) =$	.61377 .21292 −.05248 0	.60164 0 −.04186 0	.64659 2.16447 −.05984 0	.76732 5.87588 −.09296 0

Sec. 7.4   Discrete-Time-Controlled Systems   **157**

Table 7-2—*Cont.*

Problem no.	1	2	3	4
J	P1, quadratic	P1, linear	P2, linear	P2, linear
$\|J_0\|$	.92538	.91181	.14091	.22490
T_1	.51989	.54249	.51989	.54249
T_2	.54039	.54249	.54039	.57439
T_3	.54179	.54249	.54959	.60088
T_4	.54379	.54249	.55559	.61868
T_5	.54039	.54249	.55839	.62728
$\delta T_j \%$	5	0	7	13.5
C	.25	.25	—	—
$E_n(1)$	4.00	4.00	4.00	4.00
$E_n(2)$	4.00	4.00	3.76	2.88
$E_n(3)$	4.00	4.00	3.62	2.68
$E_n(4)$	4.00	4.00	3.56	2.60

J	= Performance index
$\mathbf{x}_0$	= Initial state vector
E_i	= Initial enrichment
N_i	= Number of iterations it took to converge
$\mathbf{u}_0(i)$	= Final value of the optimal-control vector at $\theta = \theta_i$ ($i = 1,\ldots, N - 1$) [$u_{03}(i)$ is preceded by a minus sign since Pu-239 is taken out]
$\|J_0\|$	= Absolute value of the optimal performance index
T_j ($j = 1,\ldots, 5$)	= Final lengths of the continuous periods (in flux time units, scaled by 10^3)
δT_j	= Maximum variation of T_j
C	= Specified Pu-239 output
$E_n(i)$	= Enrichment of loading i

Four different examples have been solved. The total number of continuous cycles considered was $N = 5$ with a maximum enrichment of 4%. The data used are summarized in Table 7-1, and the problems with the results are in Table 7-2. The LP and QP problems are solved using programs prepared by the Statistical Service Unit of the University of Illinois. These programs are written in a special SSUPAC language for the IBM 7094 Computer. The LP problem is solved there by the revised simplex method (see Chapter 3), and the QP problem by Wolfe's method, described in Chapter 3.

In addition, a sensitivity analysis is performed by perturbing the optimal values of the variables and checking the deviations of the performance indices from their optimal values. The results of this analysis are summarized in Fig. 7-8.

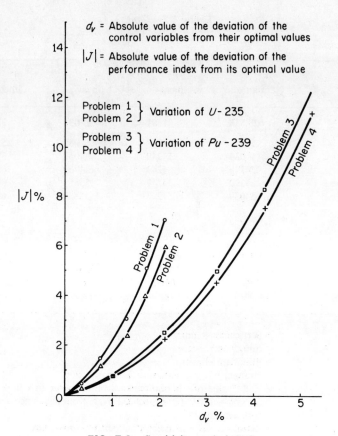

FIG. 7-8. Sensitivity analysis [16].

7.5 SAMPLED-DATA SYSTEMS WITH QUANTIZED CONTROL [12]

In many practical digital control systems the control signal is quantized; that is, the magnitude of the signal can assume only certain discrete levels. For instance, in digitally controlled systems, the control computer has a finite number of bits. This, of course, imposes quantization of its output. In this section a method of optimally controlling uniformly sampled-data systems with quantized control functions proposed by Kim and Djadjuri [12] will be discussed in conjunction with MP implementation. Uniform quantization, as illustrated in Fig. 7-9, will be assumed. The amplitude of the quantized control signal u is limited: $-u_m \leqslant u \leqslant u_m$.

Sec. 7.5 Sampled-Data Systems with Quantized Control

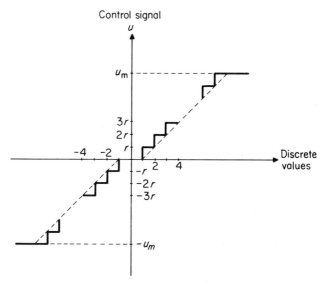

FIG. 7-9. Quantization of the control signal [12].

The system considered is described by the following set of difference state equations:

$$y(k) = \Phi(T)y(k-1) + du(k), \quad k = 1, \ldots, N \quad (7\text{-}60)$$

where $y(k) = n$-dimensional state vector at $t = t_k$
$\Phi(T) = n \times n$ state transition matrix
$T =$ Constant sampling period
$d =$ Constant n-dimensional vector
$u(k) =$ scalar, quantized control signal at $t = t_k$, which satisfies the following conditions:
  1. $u(k)$ is constant in the time interval $(k-1)T < t \leqslant kT$
  2. $u(k)$ may take only the following values (see Fig. 7-9):
  $-u_m, -u_m + r, \ldots, -2r, -r, 0, r, 2r, \ldots, u_m - r, u_m$

The initial state $y(0)$ is assumed to be specified. At any time $t = t_k$, the state $y(k)$ may be expressed as

$$y(k) = \Phi^k(T)y(0) + \sum_{i=1}^{k} \Phi^{k-i}(T)du(i) \quad (7\text{-}61)$$

To simplify the formulation of the problem, the following new variable f_k is introduced:

$$f_k = \frac{[u(k) + u_m]}{r} \quad (7\text{-}62)$$

or

$$u(k) = rf_k - u_m \tag{7-63}$$

Thus f_k is always positive and will be within the range

$$0 \leqslant f_k \leqslant \frac{2u_m}{r} \tag{7-64}$$

Furthermore, f_k can assume only integral values:

$$f_k = 0, 1, 2, \ldots, \frac{2u_m}{r} \tag{7-65}$$

Two types of performance indices are considered for this problem:

1. Minimum Quadratic Error

$$\text{Minimize } J = \sum_{k=1}^{N} \{[\mathbf{y}(k) - \mathbf{y}_d(k)]^T Q(k)[\mathbf{y}(k) - \mathbf{y}_d(k)] + wu^2(k)\} \tag{7-66}$$

where $\mathbf{y}_d(k)$ = Desired reference state vector
$Q(k) = n \times n$, symmetrical, positive-definite matrix
w = Weighting factor

2. Minimum-Time Problem

In this case, we wish to find a sequence of $u(1), \ldots, u(N)$ which drives the system from the state $\mathbf{y}(0)$ to the zero state, $\mathbf{y}(N) = 0$, in the smallest number of sampling periods.

The two problems will be discussed in detail in the following.

THE MINIMUM-QUADRATIC-ERROR PROBLEM

In this problem the number of control sampling periods N is given. To transform the problem into an integer-programming problem, the variable f_k, satisfying Eq. (7-62) and the constraint of Eq. (7-64), is considered as the control input.

Substitution of Eq. (7-63) into Eq. (7-61) results in

$$\mathbf{y}(k) = \Phi^k \mathbf{y}(0) - \sum_{i=1}^{k} \Phi^{k-i} \mathbf{d} u_m + \sum_{i=1}^{k} \Phi^{k-i} \mathbf{d} r f_i \tag{7-67}$$

The performance index can be expressed in terms of $f_1, f_2, \ldots, f_N$ by substituting Eqs. (7-67) and (7-62) into Eq. (7-66):

$$\begin{aligned}
J = &\; z_0 + z_1 f_1 + z_2 f_2 + \cdots + z_N f_N \\
&+ (z_1 + z_{11} f_1 + z_{12} f_2 + \cdots + z_{1N} f_N) f_1 \\
&\; \vdots \\
&+ (z_N + z_{N1} f_1 + z_{N2} f_2 + \cdots + z_{NN} f_N) f_N
\end{aligned} \tag{7-68}$$

where $z_{k1} = z_{1k}$, since the $Q(k)$'s are symmetrical. J is positive definite since w is nonnegative, and the $Q(k)$'s are positive definite. In matrix form the performance index can be written

$$J = z_0 + 2Z_1^T F + F^T Z_2 F \qquad (7\text{-}69)$$

where

$$F = \begin{bmatrix} f_1 \\ \cdot \\ \cdot \\ \cdot \\ f_N \end{bmatrix} \qquad Z_1 = \begin{bmatrix} z_1 \\ \cdot \\ \cdot \\ \cdot \\ z_N \end{bmatrix} \qquad Z_2 = \begin{bmatrix} z_{11} & z_{12} & \cdots & z_{1N} \\ \cdot & & & \\ \cdot & & & \\ \cdot & & & \\ z_{N1} & z_{N2} & \cdots & z_{NN} \end{bmatrix} \qquad (7\text{-}70)$$

Then the problem is to find a nonnegative sequence of control input $(f_1, f_2, \ldots, f_N)$, subject to the constraint of Eq. (7-63), which minimizes the performance index in Eq. (7-69).

The method derived in this section can be considered an eigenvector problem. In this method the optimal solution without any constraint is determined first, and from this unconstrained solution the integer solution, satisfying the constraint of Eq. (7-63), is derived.

It is convenient to describe the procedure geometrically and to illustrate it in two dimensions as in Fig. 7-10. Let F_0 represent the optimal solution without any constraint. The constraint of Eq. (7-63) is represented by an N-dimensional cube with its edge $= 2u_m/r$. In two dimensions the cube is shown by the square $ABCD$ in Figs. 7-10(a) and (b). The dimension of the cube is determined by the number of the total sampling period N.

F_0 must satisfy N linear algebraic equations,

$$\frac{\partial J}{\partial f_i} = 0 \qquad (7\text{-}71)$$

or

$$z_i + z_{i1} f_1 + \cdots + z_{iN} f_N = 0 \qquad \text{for } i = 1, 2, \ldots, N$$

In Fig. 7-10(a) F_0 lies inside the constraint cube, while in Fig. 7-10(b) F_0 lies outside the constraint cube.

Assume that the integer points satisfying Eq. (6-63) and (6-64) are lattice points denoted by dots in Fig. 7-10. Consider the functional $J(f_1, \ldots, f_N) = a_i$, where the a_i's are constant. The functional takes the form of a family of ellipsoids with point F_0 as its center. Then the problem can be regarded as finding the smallest ellipsoid which passes a lattice point.

The ellipsoid is determined as follows. Condiser an arbitrary ellipsoid $J = a_1$ denoted as ellipsoid I. The point on the ellipsoid with the largest distance from F_0 is point E_1 which lies in its largest axis, line j, in Figs. 7-10(a) and (b). The direction of line j is the same as that of the eigenvector corresponding to the smallest eigenvalue of matrix Z_2. Let Y_N denote the eigenvector. Let cube 1 be the N-dimensional cube, which has F_0 as its center; its

162 *Linear Discrete-Time Systems* Ch. 7

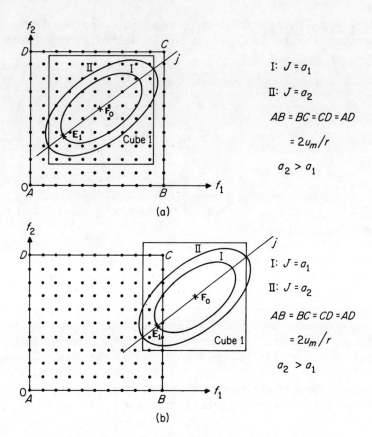

FIG. 7-10. Geometric illustration of the algorithm [12]: (a) F_0 inside the constraint cube; (b) F_0 outside the constraint cube.

rib is the distance between E_1 and F_0. Then consider the lattice points, which lie inside cube 1 and also inside the constraint cube. Let point F_{S1} be one of those points which has the smallest performance index denoted by J_{F1}. If F_{S1} lies inside or on ellipsoid I, then $J_{F1} \leqslant J_{E1}$, the performance index at point E_1. Then F_{S1} is the integer optimal solution. If $J_{F1} > J_{E1}$, point F_{S1} is outside the ellipsoid and is not the solution. Enlarge the ellipsoid, i.e., take a larger value a_2, and repeat the procedure mentioned above to get point E_2, cube 2, and point F_{S2} and then compare their corresponding J_{F2}'s and J_{E2}'s. If J_{F2} is still larger than J_{E2}, repeat it until a lattice point F_{Sk} is found (after k repetitions) whose corresponding performance index $J_{Fk} \leqslant J_{Ek}$. Then point F_{Sk} is the solution.

The algorithm can be summarized in the following steps.

Sec. 7.5   Sampled-Data Systems with Quantized Control   **163**

Step 1

Formulate the problem as mentioned above so that the performance index is as stated in Eq. (7-68).

Step 2

Find $\mathbf{F}_0$, the optimal solution without any constraint, by solving the N linear algebraic equations, Eq. (7-71). For small N, Cramer's rule can be applied here. For large N, Crout's algorithm is useful [20].

Step 3

Determine $\mathbf{Y}_N$, the eigenvector corresponding to the smallest eigenvalue of matrix Z_2. For $N \leqslant 3$ the direct method can be applied here. However, the iterative method [20] should be applied for large N. In this method, $\mathbf{Y}_N$ is determined as the eigenvector corresponding to the largest eigenvalue of Z_2^{-1}. Matrix Z_2^{-1} itself can be determined by applying Crout's algorithm.

Step 4

Let the distance between $\mathbf{F}_1$ and $\mathbf{E}_1$ be kd, where d is an arbitrary number. Starting with $k = 1$, find point $\mathbf{E}_1$, where the direction of $F_0 E_1$ coincides with that of $\mathbf{Y}_N$ and the distance $F_0 E_1$ is d. Compute also the performance index at point $\mathbf{E}_1$, that is, J_{E1}.

Step 5

Determine the coordinates of the lattice points which lie inside cube 1 (whose edge is d) and also inside the constraint cube. Determine, among those lattice points, point F_{S1} which has the smallest performance index J_{F1}. This can be obtained by direct computations and comparisons for small N or by applying dynamic programming for large N.

Step 6

Compare J_{F1} and J_{E1}. If $J_{F1} \leqslant J_{E1}$, then F_{S1} is the optimal solution. If $J_{F1} > J_{E1}$, repeat Steps 4, 5, and 6 with $k = 2, 3, \ldots$ until point F_{Sk} is found whose corresponding $J_{Fk} \leqslant J_{Ek}$, where F_{Sk} is then the optimal solution.

ILLUSTRATIVE EXAMPLE

The transfer function of the linear plant considered is given as

$$G(s) = \frac{1}{s(s+1)} \qquad (7\text{-}72)$$

The state vector is defined as

$$\mathbf{y}(t) = \begin{bmatrix} y(t) \\ \dot{y}(t) \end{bmatrix} \tag{7-73}$$

The sampling period T is 1 sec. The fundamental and input matrices are computed:

$$\Phi(T) = \begin{bmatrix} 1 & .6321 \\ 0 & .3679 \end{bmatrix}$$

and

$$\mathbf{d} = \begin{bmatrix} .3679 \\ .6321 \end{bmatrix} \tag{7-74}$$

The control function can take on integer values as follows:

$$u_k = \text{Integer}, \quad -10 \leqslant u_k \leqslant 10$$

The desired state $\mathbf{y}_d(k)$ is the zero state when w is zero and

$$Q(k) = \begin{bmatrix} 1 & 0 \\ 0 & 1 \end{bmatrix} \tag{7-75}$$

The initial state is given as

$$\mathbf{y}(0) = \begin{bmatrix} 5.0 \\ 0 \end{bmatrix} \tag{7-76}$$

By substituting Eq. (7-75) into Eq. (7-66), the performance index is

$$J = \sum_{k=1}^{N} \mathbf{y}(k)^T \mathbf{y}(k)$$

or

$$J = \sum_{k=1}^{N} [y(k)^2 + \dot{y}(k)^2] \tag{7-77}$$

Introduce new variables:

$$f_k = u_k + 10$$

or

$$u_k = f_k - 10, \quad k = 1, 2, \ldots, N \tag{7-78}$$

It is obvious that f_k can take on only positive integer values as follows:

$$0, 1, 2, \ldots, 19, 20 \tag{7-79}$$

The problem is to determine a sequence of $f_1, \ldots, f_N$ which minimizes the performance index, Eq. (7-77), subject to the constraint of Eq. (7-79).

For $N = 2$

Substitution of Eqs. (7-74) and (7-78) into Eq. (7-67) results in

$$\mathbf{y}(1) = \begin{bmatrix} 1.3212 + .3679 f_1 \\ -6.3212 + .6321 f_1 \end{bmatrix}$$

$$\mathbf{y}(2) = \begin{bmatrix} -6.3534 + .7675 f_1 + .3679 f_2 \\ -8.6466 + .2325 f_1 + .6321 f_2 \end{bmatrix} \tag{7-80}$$

Sec. 7.5 Sampled-Data Systems with Quantized Control

From Eqs. (7-77) and (7-80) the performance index is

$$J = 156.8328 - 10.3964f_1 - 7.8030f_2 \\ + (-10.3964 + 1.1780f_1 + .4293f_2)f_1 \\ + (-7.8030 + .4293f_1 + .5349f_2)f_2 \quad (7\text{-}81)$$

Point F_0 (4.9600, 10.6065) represents the optimal solution without any constraint, which satisfies the following two equations:

$$\begin{aligned} 1.1780f_1 + .4293f_2 &= 10.3964 \\ .4293f_1 + .5349f_2 &= 7.8030 \end{aligned} \quad (7\text{-}82)$$

The eigenvalues of matrix Z_2 are determined by the direct method, as follows. The determinant of $Z_2 - \gamma I$ is zero; that is,

$$\begin{vmatrix} 1.1790 - \gamma & .4293 \\ .4293 & .5349 - \gamma \end{vmatrix} = 0 \quad (7\text{-}83)$$

The solutions of Eq. (7-83) are $\gamma_1 = 1.3928$ and $\gamma_2 = .3201$. Assume that $\mathbf{Y}_2$ is the eigenvector corresponding to γ_2, which is determined from the following equation:

$$Z_2 \mathbf{Y}_2 = \gamma_2 \mathbf{Y}_2 \quad (7\text{-}84)$$

Then the eigenvector is

$$\mathbf{Y}_2 = \begin{bmatrix} 1.0 \\ -1.9983 \end{bmatrix}$$

Let e_1 and e_2 be the coordinates of point E_1, where the distance PE_1 is $k\sqrt{2}$ (in this case d is $\sqrt{2}$). Starting with $k = 1$, the distance becomes 2 and

$$\mathbf{E}_1 = \begin{bmatrix} e_1 \\ e_2 \end{bmatrix} = \frac{\mathbf{F}_0 + d\mathbf{Y}_2}{\|\mathbf{Y}_2\|} \quad (7\text{-}85)$$

where $\|\mathbf{Y}_2\|$ is the length of vector $\mathbf{Y}_2$. Then the point is E_1 (5.5928, 9.3418) and its corresponding performance index J_{E1} is 23.1450. For the lattice points inside square 1, the possible values of f_1 are 4.0, 5.0, and 6.0, and for f_2 they are 10.0, 11.0, and 12.0. Among these points, point F_{S1} (5.0, 11.0) has the smallest performance index; that is, $J_{F1} = 22.6031$. Point F_{S1} is the optimal integer solution since $J_{F1} < J_{E1}$. The optimal controls are $f_1 = 5.0, f_2 = 11.0$ or $u_1 = -5.0, u_2 = 1.0$, with a minimum performance index of $J_2 = 22.6031$. The corresponding state vectors are

$$\mathbf{y}(1) = \begin{bmatrix} 3.1606 \\ -3.1606 \end{bmatrix} \quad \text{and} \quad \mathbf{y}(2) = \begin{bmatrix} 1.5306 \\ -.5306 \end{bmatrix} \quad (7\text{-}86)$$

For $N = 5$

The state vectors $\mathbf{y}(1)$ and $\mathbf{y}(2)$ are written in Eq. (7-80). Substitutions of Eqs. (7-74) and (7-78) into Eq. (7-67) results in

$$\mathbf{y}(3) = \begin{bmatrix} -15.4979 + .9145f_1 + .7675f_2 + .3679f_3 \\ -9.5021 + .0855f_1 + .2325f_2 + .6321f_3 \end{bmatrix}$$

$$\mathbf{y}(4) = \begin{bmatrix} -25.1832 + .9685f_1 + .9145f_2 + .7675f_3 + .3679f_4 \\ -9.8168 + .0315f_1 + .0855f_2 + .2325f_3 + .6321f_4 \end{bmatrix} \quad (7\text{-}87)$$

$$y(5) = \begin{bmatrix} -35.0674 + .9884f_1 + .9685f_2 + .9145f_3 + .7675f_4 + .3679f_5 \\ -9.9326 + .0116f_1 + .0315f_2 + .0855f_3 + .2325f_4 + .6321f_5 \end{bmatrix}$$

From Eqs. (7-77), (7-80), and (7-87), the performance index becomes

$$\begin{aligned} J = {} & 2546.2471 - 84.8572f_1 - 80.0515f_2 - 66.2348f_3 - 44.6922f_4 - 19.1792f_5 \\ & + (-84.8572 + 3.9377f_1 + 2.9971f_2 + 2.0460f_3 + 1.1375f_4 + .3709f_5)f_1 \\ & + (-80.0515 + 2.9971f_1 + 2.9606f_2 + 2.0394f_3 + 1.411f_4 + .3762f_5)f_2 \\ & + (-66.2348 + 2.0460f_1 + 2.0394f_2 + 2.0215f_3 + 1.1510f_4 + .3905f_5)f_3 \\ & + (-44.6922 + 1.1375f_1 + 1.1411f_2 + 1.1510f_3 + 1.1780f_4 + .4293f_5)f_4 \\ & + (-19.1792 + .3709f_1 + .3762f_2 + .3905f_3 + .4293f_4 + .5349f_5)f_5 \end{aligned}$$

(7-88)

Point F_0 satisfies the following five equations:

$$\frac{\partial J}{\partial f_i} = 0 \quad \text{for } i = 1, \ldots, 5 \tag{7-89}$$

When Crout's algorithm is applied, the optimal solution without any constraint is represented by point F_0 (4.7388, 10.1746, 10.0601, 10.0250, 10.0230). The inverse of matrix Z_2, also determined by Crout's algorithm, is

$$Z_2^{-1} = \begin{bmatrix} 1.1075 & -1.1437 & .0251 & .0105 & .0096 \\ -1.1437 & 2.2896 & -1.1676 & .0204 & .0188 \\ .0251 & -1.1676 & 2.2941 & -1.1554 & .0563 \\ .0105 & .0204 & -1.1554 & 2.3301 & -1.0483 \\ .0096 & .0188 & .0563 & -1.0483 & 2.6498 \end{bmatrix} \tag{7-90}$$

The eigenvector Y_5 is determined by the iterative method [20]. The initial iteration is

$$V_0 = W_0 = (1 \ 0 \ 0 \ 0 \ 0)^T$$

The nth iteration is related to its next iteration by

$$W_{n+1} = Z_2^{-1}V_n = g_{n+1}V_{n+1} \tag{7-91}$$

where the first element of V_k is 1.0. The matrix multiplication in Eq. (7-91) is repeated until the difference between two consecutive g factors satisfies

$$\left| \frac{g_{n+1} - g_n}{g_n} \right| \leqslant .0001 \tag{7-92}$$

This happens after 32 iterations. A.C. Aitken's extrapolation [20] can be applied here to improve the accuracy of the iterations, as follows. Let $v_{i,n-1}$, $v_{i,n}$, and $v_{i,n+1}$ be the ith component of V_{n-1}, V_n, and V_{n+1}, respectively. Then the ith component of Y_5 is

$$t_i = \frac{v_{i,n-1} \cdot v_{i,n+1} - v_{i,n}}{v_{i,n-1} - 2v_{i,n} + v_{i,n+1}} \tag{7-93}$$

After Eq. (7-93) is applied, the eigenvector is obtained:

$$Y_5 = (1.0, -2.7107, 3.6311, -3.3858, 2.2701)^T \tag{7-94}$$

Starting with $k = 1$, while the distance F_0E_1 is 1.0, the coordinates of point E are 4.9007, 9.7357, 10.6480, 9.4768, and 10.3905, whose performance index is $J_{E1} = 23.2670$. For the lattice points inside cube 1 and also inside the constraint cube, the possible values of f_1 are 4.0, 5.0, while those of f_2, f_3, f_4, and f_5 are 10.0, 11.0. By direct computations and comparisons, it is found that point F_{S1} (5.0, 10.0, 10.0, 10.0, 10.0) has the smallest performance index, $J_{F1} = 23.1048$. Since $J_{F1} < J_{E1}$, the optimal integer solution is $u_1 = -5.0, u_2 = u_3 = u_4 = u_5 = 0$, with a minimum performance index of $J_5 = 23.1048$. The corresponding state vectors are

$$\mathbf{y}(1) = \begin{bmatrix} 3.1606 \\ -3.1606 \end{bmatrix}, \quad \mathbf{y}(2) = \begin{bmatrix} 1.1627 \\ -1.1627 \end{bmatrix}, \quad \mathbf{y}(3) = \begin{bmatrix} .4277 \\ -.4277 \end{bmatrix},$$

$$\mathbf{y}(4) = \begin{bmatrix} .1574 \\ -.1574 \end{bmatrix}, \quad \mathbf{y}(5) = \begin{bmatrix} .0579 \\ -.0579 \end{bmatrix}$$

(7-95)

THE MINIMUM-TIME PROBLEM

In this problem the least number of sampling periods N and its corresponding sequence of controls which will drive the plant within zero state should be determined. To transform the problem into a mixed-integer LP problem, the control variable $f(t)$, which satisfies Eq. (7-63) and the constraint of Eq. (7-64), is considered.

Introduce nonnegative slack variables $(q_1, q_2, \ldots, q_N)$ into the inequality in Eq. (7-64) to obtain

$$f_k + q_k = \frac{2u_m}{r} \quad \text{for } k = 1, 2, \ldots, N \tag{7-96}$$

It is obvious that the q_k's are nonnegative integer variables. Substitution of Eq. (7-63) into Eq. (7-61) gives

$$\mathbf{y}(N) = \begin{bmatrix} y_1 \\ \cdot \\ \cdot \\ \cdot \\ y_n \end{bmatrix}$$

$$= \Phi^N \mathbf{y}(0) - \sum_{m=1}^{N} \Phi^{N-m} \mathbf{d}u_m + \sum_{m=1}^{N} \Phi^{N-m} \mathbf{d}rf_m \tag{7-97}$$

where n is the order of the system. If the matrix multiplication is performed on Eq. (7-97), the elements of $\mathbf{y}(N)$ are

$$y_1 = b_{1,0} + b_{1,1}f_1 + \cdots + b_{1,N}f_N$$

$$\vdots \tag{7-98}$$

$$y_n = b_{n,0} + b_{n,1}f_1 + \cdots + b_{1,N}f_N$$

It is desired that the elements of $\mathbf{y}(N)$ satisfy

$$y_k \leqslant y_0 \quad \text{for } k = 1, 2, \ldots, n \tag{7-99}$$

where y_0 is a given number. Introduce slack variables $(g_1, g_2, \ldots, g_n)$ which satisfy

$$g_k = y_k + y_0 \qquad (7\text{-}100)$$

and

$$0 \leqslant g_k \leqslant 2y_0 \qquad \text{for } k = 1, 2, \ldots, n \qquad (7\text{-}101)$$

With introduction of other nonnegative variables $(p_1, p_2, \ldots, p_n)$, the inequality in Eq. (7-101) can be rewritten as

$$g_k + p_k = 2y_0 \qquad \text{for } k = 1, 2, \ldots, n \qquad (7\text{-}102)$$

Substitution of Eq. (7-100) into Eq. (7-102) results in

$$b_{k,1}f_1 + \cdots + b_{k,N}f_N - g_k = -b_{k,0} - y_0 \qquad \text{for } k = 1, 2, \ldots, n \qquad (7\text{-}103)$$

Since all constraints are linear in f_k's, the problem can be considered as a LP problem. To get an objective function, introduce nonnegative artificial variables $(a_1, a_2, \ldots, a_n)$ into Eq. (7-103). If $b_{k,0} + y_0 < 0$, rewrite Eq. (7-103) as

$$a_k + b_{k,1}f_1 + \cdots + b_{k,N}f_N - g_k = -b_{k,0} - y_0$$
$$\text{for } k = 1, 2, \ldots, n \qquad (7\text{-}104)$$

If $b_{k,0} + y_0 > 0$, rewrite Eq. (7-103) as

$$a_k - b_{k,1}f_1 - \cdots - b_{k,N}f_N + g_k = b_{k,0} + y_0$$
$$\text{for } k = 1, 2, \ldots, n \qquad (7\text{-}105)$$

Consider the functional

$$Z = -a_1 - a_2 - \cdots - a_n \qquad (7\text{-}106)$$

Since the artificial variables must be zero for any feasible solution at the end of computation, Z can be considered as an objective function to be maximized. The problem is a mixed-integer one, that is, to find the smallest N and its corresponding nonnegative integer sequence $(f_1, f_2, \ldots, f_N; q_1, q_2, \ldots, q_N)$ and nonnegative sequence $(g_1, g_2, \ldots, g_n; p_1, p_2, \ldots, p_n)$ which will maximize Z into zero, subject to the constraints, Eqs. (7-96), (7-102), (7-104), and (7-105). To solve the problem, two phases of operations are performed [21]. The first phase is to solve the noninteger LP problem to determine the minimum N and its control sequence without the integer constraints on f_k and q_k, which will drive Z into zero. This is equivalent to finding a basic feasible solution in LP (see Chap. 3)

The second phase, based on the solution of the first phase, is to find a sequence of integer controls which will give $y(N)$ satisfying the constraint of Eq. (7-99).

Since N is unknown, an iterative procedure must be used. At first, an attempt to find a basic feasible solution is made for $N = 1$. If no basic feasible solutions exist for $N = 1$, an attempt is made for $N = 2$. This process is continued until the smallest N is found for which the constraints, Eqs. (7-96),

(7-102), and (7-103), are satisfied. This means that the first phase ends. With the resulting basic feasible solution, the second phase is initiated.

METHOD OF SOLUTION THROUGH THE QUADRATIC AND EIGENVECTOR PROBLEM

In this method the minimum-time problem is considered and solved as a quadratic error problem. It was mentioned in the previous section that the state of the plant $\mathbf{y}(N)$ should satisfy the constraint of Eq. (7-99). This is equivalent to the following constraint:

$$\mathbf{y}(N)^T\mathbf{y}(N) \leqslant \beta \tag{7-107}$$

where β is a specified number. A new performance index is defined:

$$J = \mathbf{y}(N)^T\mathbf{y}(N) \tag{7-108}$$

Then the constraint of Eq. (7-107) becomes

$$J \geqslant \beta \tag{7-109}$$

The problem can be solved by the following iterative method. Start with $N = 1$ and determine the optimal solution, subject to the constraint of Eq. (7-64), which will minimize the performance index, Eq. (7-108). If the minimum performance index is satisfied in Eq. (7-109), then the problem is solved. If not, repeat the process until a value of N is found for which the corresponding minimum performance index satisfies inequality (7-109).

7.6 CONCLUDING REMARKS

It was demonstrated in several case studies in this chapter that LP and QP can be effectively applied to the solution of linear discrete-time optimal-control problems. However, in some cases, especially in highly dimensioned systems, we obtain an excessively large tableau for the LP or QP algorithm. This can lead to some ill-conditioning effects, where inversion of matrices close to being singular is attempted. These ill-conditioning effects were recently investigated by Polak [22], who proposed some alternative algorithms which avoid this type of difficulty. Further investigation in implementing the new algorithms in a wider variety of optimal-control problems constitutes an interesting problem for future research.

REFERENCES

1. B. C. Kuo, *Analysis and Synthesis of Sampled-Data Control Systems*, Prentice-Hall, Englewood Cliffs, N.J., 1963.

2. B. H. Whalen, "Linear Programming for Optimal Control," Ph. D. Thesis, University of California, Berkeley, 1962.
3. L. A. Zadeh and B. H. Whalen, "On Optimal Control and Linear Programming," *IRE Trans. Automatic Control*, **AC-7**, pp. 45–46, 1962.
4. A. I. Propoi, "Use of Linear Programming Methods for Synthesizing Sampled-Data Automatic Systems," *Automation Remote Control*, **24**, pp. 912–920, 1963.
5. K. A. Fegley, "Designing Sampled-Data Control Systems by Linear Programming," *IEEE Trans. Appl. Ind.*, **83**, pp. 198–200, 1964.
6. H. C. Torng, "Optimization of Discrete Control Systems Through Linear Programming," *J. Franklin Inst.*, **277**, pp. 28–44, 1964.
7. G. Porcelli and K. A. Fegley, "Linear Programming Design of Digitally Compensated Systems," *JACC*, **1964** pp. 412–421, Stanford, Cal.
8. K. A. Fegley and M. I. Hsu, "Optimum Discrete Control by Linear Programming," *IEEE Trans. Automatic Control*, **AC-10**, pp. 114–115, 1965.
9. G. Porcelli and K. A. Fegley, "Optimal Design of Digitally Compensated Systems by Quadratic Programming," *J. Franklin Inst.*, **282**, pp. 303–317, 1966.
10. M. Kim, "On Optimum Control of Discrete Systems: I. Theoretical Development," *ISA Trans.*, **5**, pp. 93–98, 1966.
11. M. D. Canon and J. H. Eaton, "A New Algorithm for a Class of Quadratic Programming Problems with Application to Control," *J. SIAM Control*, **4**, pp. 34–45, 1966.
12. M. Kim and K. Djadjuri, "Optimum Sampled Data Control Systems with Quantized Control Function," *ISA Trans.*, **6**, pp. 65–73, 1967.
13. G. N. T. Lack and M. Enns, "Optimal Control Trajectories with Minimax Objective Functions by Linear Programming," *IEEE Trans. Automatic Control*, **AC-12**, pp. 749–752, 1967.
14. S. Blum and K. A. Fegley, "A Quadratic Programming Solution of the Minimum Energy Control Proglem," *IEEE Trans.*, **AC-13** *Automatic Control*, pp. 206–207, 1968.
15. D. Tabak, "Application of Mathematical Programming in the Design of Optimal Control Systems," Ph.D. Thesis, University of Illinois, Urbana, 1967.
16. D. Tabak, "Optimization of Nuclear Reactor Fuel Recycle via Linear and Quadratic Programming," *IEEE Trans. Nuclear Science*, **NS-15**, pp. 60–64, 1968.
17. L. Cutler, *Product Form Quadratic Programming Code RSQPF4*, The RAND Corporation, Santa Monica, Cal., 1965.
18. D. Tabak and B. C. Kuo, "Application of Mathematical Programming in the Design of Optimal Control Systems," *Intern. J. Control*, **10**, pp. 545–552, 1969.
19. D. Tabak, "A Direct and Nonlinear Programming Approach to the Optimal Nuclear Reactor Shutdown Control," *IEEE Trans. Nuclear Science*, **NS-15**, pp. 57–59, 1968.

20. S. H. Crandall, *Engineering Analysis*, McGraw-Hill, New York, 1956, pp. 26–47 and 91–98.
21. R. E. Gomory, "An Algorithm for Integer Solutions to Linear Programs," in *Recent Advances in Mathematical Programming* (R. L. Graves and P. Wolfe, eds.), McGraw-Hill, New York, 1963, pp. 269–302.
22. E. Polak, "On the Removal of Ill-Conditioning Effects in the Computation of Optimal Controls," *Automatica*, **5**, pp. 607–614, 1969.

8

NONLINEAR DISCRETE-TIME SYSTEMS

8.1 INTRODUCTION

After the formulation of the maximum principle for optimal control of continuous time systems was reported [1], considerable work was done in establishing a parallel version of the maximum principle for discrete-time systems, i.e., the so-called distrete (or digitized) maximum principle. The extension has been made in references 2, part III, and 3–9, to mention only a few. In almost all the references mentioned, necessary conditions for optimality have been derived. However, in every case one is faced with solving a difficult two-point boundary-value problem.

It has been found that, at present, one of the most efficient ways of getting computational results in the design of optimal discrete-time control systems, and especially in the case of nonlinear problems, is by the use of mathematical programming. In the case of linear discrete-time systems, we can use the dynamic-programming approach for low-dimensioned systems, as was done by Tou [10] for uniform sampling. A suboptimal solution for nonuniform sampling has been proposed by Brockstein and Kuo [11]. Although mathematical programming, in the nonlinear case, has its own computational difficulties, it is still a method by which numerical results can be obtained with relatively little complexity. Similar ideas have also been expressed by Rosen [12].

The dynamic system under consideration is governed by the following set of state difference equations:

$$\mathbf{y}(i+1) = \mathbf{f}[\mathbf{y}(i), \mathbf{u}(i), i], \quad i = 0, 1, \ldots, N-1 \quad (8\text{-}1)$$

where $\mathbf{y}(i) = n$-dimensional state vector at the discrete-time instant $t = t_i$
$\mathbf{u}(i) = m$-dimensional control vector at $t = t_i$
$\mathbf{f} =$ Nonlinear, time-varying, n-dimensional vector function
$N =$ Maximum number of discrete-time intervals considered; the intervals are, in general, unequal

At each discrete sampling instant, the system is subjected to an additional set of equality and inequality constraints:

$$h_j[\mathbf{y}(i), \mathbf{u}(i)] = 0, \quad j = 1, \ldots, p \quad (8\text{-}2)$$

$$g_k[\mathbf{y}(i), \mathbf{u}(i)] \geq 0, \quad k = 1, \ldots, q \quad (8\text{-}3)$$

where h_j and g_k are generally nonlinear functions.

It will be assumed that the initial state vector is specified:

$$\mathbf{y}(0) = \mathbf{y}_0 \quad (8\text{-}4)$$

The general optimal-control problem for the discrete-time system under consideration can be formulated as follows:

Bring the system described by Eq. (8-1) from the initial state $\mathbf{y}_0$ into a target area described by

$$\mathbf{a}[\mathbf{y}(N)] \geq 0 \quad (8\text{-}5)$$

where $\mathbf{a}$ is an l-dimensional, nonlinear, vector function, so that the following performance index is minimized (or maximized):

$$J = \sum_{i=1}^{N} F[\mathbf{y}(i), \mathbf{u}(i-1), i] \quad (8\text{-}6)$$

where F is a nonlinear time-varying function subject to the constraints in Eqs. (8-2) and (8-3).

It is obvious that an optimal-control problem for a discrete-time system with a finite number of sampling intervals N is also a MP problem of finite dimensions. In analogy with the formulation in Eq. (4-26), we can reformulate the optimal-control problem stated above as follows:

$$\min \text{ (or max)} \left\{ \sum_{i=1}^{N} F[\mathbf{y}(i), \mathbf{u}(i-1), i] \,\middle|\, \mathbf{a}[\mathbf{y}(N)] \geq 0; \right.$$
$$g_k[\mathbf{y}(i), \mathbf{u}(i-1)] \geq 0, k = 1, \ldots, q, h_j[\mathbf{y}(i),$$
$$\mathbf{u}(i-1)] = 0, j = 1, \ldots, p; \mathbf{h}_{p+i+1}$$
$$\left. = \mathbf{y}(i+1) - \mathbf{f}[\mathbf{y}(i), \mathbf{u}(i-1), i] = 0, i = 0, 1, \ldots, N-1 \right\} \quad (8\text{-}7)$$

Since the functions $\mathbf{f}$ are assumed to be nonlinear, one obviously has a NLP problem. In the most general case all other functions—F, $\mathbf{a}$, $\mathbf{g}$, $\mathbf{h}$—

are assumed to be nonlinear. The sampling times may be uniform or non-uniform, known or unknown.

The Kuhn-Tucker theorem [13] can be applied to the MP problem in Eq. (8-7) (see Sec. 2.6). To conform with the previous formulation given in Sec. 2.6, the following set of functions is introduced:

$$G_1 \equiv -a_1$$
$$\vdots$$
$$G_l \equiv -a_l$$
$$G_{l+1} \equiv -g_1$$
$$\vdots$$
$$G_{l+q} \equiv -g_q$$

It will be assumed that the functions F, G_i ($i = 1, \ldots, l+q$), and $h_j(j = 1, \ldots, p + nN)$ are convex. Furthermore, to be more specific, it will be assumed that we have a minimization problem:

$$\min \left\{ \sum_{i=1}^{N} F[\mathbf{y}(i), \mathbf{u}(i-1), i] \,|\, G_j[\mathbf{y}(i), \mathbf{u}(i-1)] \leqslant 0, j = 1, \ldots, \right.$$
$$\left. l + q; h_k[\mathbf{y}(i), \mathbf{u}(i-1), i] = 0, k = 1, \ldots, p + nN \right\} \quad (8\text{-}8)$$

Applying Kuhn-Tucker's theorem, we can form a Hamiltonian function, analogous to the Lagrange function of Eq. (2-23):

$$H[\mathbf{y}(i), \mathbf{u}(i-1), i] = \sum_{i=1}^{N} F[\mathbf{y}(i), \mathbf{u}(i-1), i]$$
$$+ \sum_{j=1}^{M} \mu_j G_j[\mathbf{y}(i), \mathbf{u}(i-1)] \quad (8\text{-}9)$$
$$+ \sum_{k=1}^{Q} \lambda_k h_k[\mathbf{y}(i), \mathbf{u}(i-1), i], \quad i = 1, \ldots, N$$

where $M = l + q$
$Q = p + nN$

The μ_j's are the Kuhn-Tucker multipliers, and the λ_k's are the Lagrange multipliers.

If $\mathbf{y}^o(i)$ and $\mathbf{u}^o(i-1)$ ($i = 1, \ldots, N$) denote the optimal state and control vectors, respectively, one may use the right-hand side of the Kuhn-Tucker inequality in Eq. (2-24) to obtain

$$H[\mathbf{y}^o(i), \mathbf{u}^o(i), \mu^o, \lambda^o] \leqslant H[\mathbf{y}(i), \mathbf{u}(i), \mu^o, \lambda^o] \quad (8\text{-}10)$$

where μ^o and λ^o are the optimal vectors of the Kuhn-Tucker coefficients μ_k and the Lagrange multipliers λ_k, respectively.

The inequality in Eq. (8-10) can be regarded as a sort of minimum-principle formulation of an optimal-control problem for a discrete-time system. Similar results have been reported previously by Rosen [12], Pearson and Sridhar [14], and Mangasarian and Fromovitz [15]. Necessary conditions for a general optimization problem have been derived by Canon, Cullum, and Polak [16]. As particular cases of these conditions, the authors have derived the Kuhn-Tucker theorem and the discrete maximum principle. These results support the point made in Chapter 4 that an optimal-control problem is actually a MP problem. Perhaps we could regard mathematical programming as a computational realization of the necessary conditions of optimality given by the maximum principle [1].

8.2 EXAMPLE: A NONLINEAR SECOND-ORDER SYSTEM [17, 18]

A second-order system with one control variable is considered in the following example. The system is described by the following difference state equations:

$$y_1(i+1) - y_1(i) - T_{i+1}[-y_1^2(i) + y_2(i) + u(i)] = 0$$
$$y_2(i+1) - y_2(i) - T_{i+1}y_1(i) = 0, \quad (8\text{-}11)$$
$$i = 0, 1, \ldots, N-1$$

where $y_j(i) = j$th component of the state vector at the discrete time t_i,

$$T_j = t_j - t_{j-1}$$

The initial state of the system is

$$\mathbf{y}(0) = \begin{bmatrix} 0 \\ 1 \end{bmatrix}$$

The problem is to bring the initial state into the target area:

$$[y_1(N) - 10]^2 + y_2^2(N) - 1 \leq 0 \quad (8\text{-}12)$$

so that the performance index

$$J = \sum_{i=1}^{N} [y_1^2(i) + y_2^2(i) + .1u^2(i-1)] \quad (8\text{-}13)$$

is minimized. The control variable is assumed to be bounded:

$$0 \leq u(i) \leq 1, \quad i = 0, 1, \ldots, N-1 \quad (8\text{-}14)$$

And the state variables are to be nonnegative:

$$y_j(i) \geq 0, \quad i = 1, \ldots, N \quad (8\text{-}15)$$
$$j = 1, 2$$

Notice that the target area represents a circle of radius 1 around the point (10, 0). Two computational examples are solved. In both cases the number of sampling periods is chosen to be $N = 12$. The difference between these is that in the first case, the total time

$$T = \sum_{i=1}^{N} T_i$$

is assumed free, whereas in the second case, T is fixed at 10.

The results of the free-time example are tabulated as follows:

i	T_i (sec)	t_i (sec)	$u(i-1)$	$y_1(i)$	$y_2(i)$
1	.00291	.00291	.0553	.00265	.964
2	.00196	.00487	.5000	.00375	.930
3	.00176	.00663	.5000	.00434	.898
4	.00169	.00832	.0466	.00467	.867
5	.00164	.00996	.0435	.00479	.839
6	.00162	.01158	.0419	.00473	.812
7	.00159	.01317	.0406	.00448	.786
8	.00156	.01473	.0395	.00402	.762
9	.00151	.01624	.0383	.00330	.740
10	.00144	.01768	.0370	.00218	.719
11	.00130	.01898	.0356	.00012	.699
12	13.18224	13.20122	.0338	9.26978	.683

The optimal performance index is computed to be

$$J_{\min} = 93.86$$

The time variations of y_1, y_2, and u are presented in Fig. 8-1.

For the fixed-time case, the following results are obtained:

i	T_i (sec)	t_i (sec)	$u(i-1)$	$y_1(i)$	$y_2(i)$
1	.00146	.00146	.404	.00205	.973
2	.07106	.07252	.500	.14261	.948
3	.00042	.07294	.500	.12606	.925
4	.00043	.07337	.992	.10976	.903
5	.00044	.07381	.023	.09369	.884
6	.00046	.07427	.023	.07781	.866
7	.00047	.07474	.023	.06208	.850
8	.00048	.07522	.022	.04648	.832
9	.00048	.07570	.022	.03099	.822
10	.00049	.07619	.022	.01555	.811
11	.00050	.07669	.022	.00012	.801
12	9.92503	10.00172	.022	9.36986	.776

$$J_{\min} = 96.91$$

Sec. 8.2   Example: A Nonlinear Second-Order System   **177**

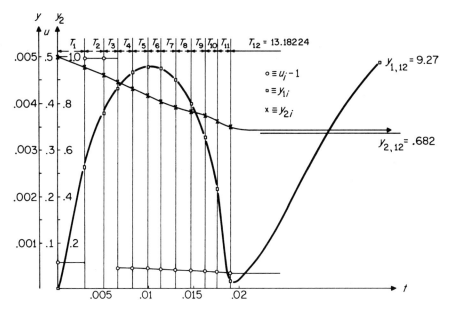

FIG. 8-1. The free-time case [18].

From these results it is noticed that a smaller performance index is obtained in the free-time case. The results of the second example are presented graphically in Fig. 8-2.

In both cases, the time variation of all the variables are similar. As we

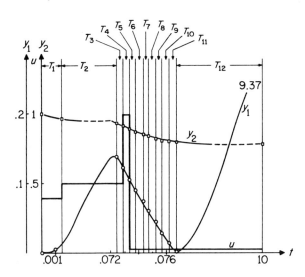

FIG. 8-2. The fixed-time case [18].

can see from Figs. 8-1 and 8-2, most of the control effort is made during the first three or four discrete intervals. After that, a very small control signal is applied. The response of y_1 exhibits an overshoot, undershoots to a very small value, and then increases directly to its target value in one long sampling interval. The response of y_2 decreases very slowly during the process down to about 68–77% of its initial value.

In both cases, the NLP problem had 48 variables and 37 constraints. The computation time for each problem was approximately 8 min on the IBM 7094 Computer. The code used in this example was the SUMT code, originated by Fiacco and McCormick [19].

8.3 AN ANTENNA-TRACKING CONTROL SYSTEM WITH A NONLINEARITY [27]

A part of an antenna-tracking control system is considered. The system consists of a separable linear and nonlinear subsystem. The dynamics of the system are represented by the following set of difference equations:

$$\mathbf{y}(i+1) = A\mathbf{y}(i) + \mathbf{b}NL[u(i)] \tag{8-16}$$

where $A = n \times n$ constant matrix
$\mathbf{b} = n$-dimensional constant vector
$u(i) =$ Scalar control variable at $t = t_i$
$NL =$ Nonlinear scalar function

The system under consideration is uniformly sampled. In view of this, we can now express the state vector at any time instant $t = t_N$ as [10]

$$\mathbf{y}(N) = A_N \mathbf{y}(0) + \sum_{i=0}^{N-1} A^{N-i-1} \mathbf{b}NL[u(i)] \tag{8-17}$$

where $\mathbf{y}(0)$ is the initial state vector, which is assumed known.

The output of the system is represented by the first component of the state vector, y_1. The purpose of the control action is to bring the output y_1 to align with a prescribed reference position y_R. This control action should be done with minimum expense of energy. Following these requirements the following performance index is formulated:

$$\text{Minimize } J = \sum_{i=1}^{N} \{[y_1(i) - y_R]^2 + Wu^2(i-1)\} \tag{8-18}$$

where W is a weighting factor and N is the maximum number of sampling instants considered.

The meaning of the first term in the performance index in Eq. (8-18) is minimization of the squared error, at any time instant considered, between the actual and the desired output position of the system. The second term represents the minimum energy requirement.

From Eq. (8-17) it follows that

$$y_1(i) = A_1^i \mathbf{y}(0) + \sum_{j=0}^{i-1} A_1^{i-j-1} \mathbf{b} NL[u(j)] \tag{8-19}$$

where A_1 represents the first row of the matrix A.

Substituting Eq. (8-19) into Eq. (8-18), we obtain

$$J = \sum_{i=1}^{N} \left\{ \left[A_1^i \mathbf{y}(0) + \sum_{j=0}^{i-1} A_1^{i-j-1} \mathbf{b} NL[u(j)] - y_R \right]^2 + W u^2(i-1) \right\} \tag{8-20}$$

It is clear that Eq. (8-20) represents a nonlinear performance index or objective function.

Additional constraints limiting the amplitude of the control signal are

$$-u_m \leqslant u(i) \leqslant u_m \tag{8-21}$$

The objective function in Eq. (8-20), along with the constraints in Eq. (8-21), forms an NLP problem with N variables and $2N$ inequality constraints.

The algorithm applied for the solution of this problem is the SUMT method [19] on the IBM 360/91 Computer System.

The actual system considered is modeled as a third-order system for which

$$A = \begin{bmatrix} -.114 & 1.000 & 0 \\ .233 & -.114 & 1 \\ -.275 & .233 & 0 \end{bmatrix} \quad \mathbf{b} = \begin{bmatrix} .179 \\ .550 \\ .115 \end{bmatrix}$$

The nonlinearity considered is of the saturation type and is modeled by a hyperbolic tangent function

$$NL(u) = S \tanh \frac{u}{S} \tag{8-22}$$

where S is the output of the nonlinearity at saturation, or in this case, the output when $u \to \infty$. On many occasions, a saturation is modeled using sharp corners at the passage from the linear region to saturation. In many practical systems, such as in servo amplifiers, this passage is rather smooth, and the function in Eq. (8-22) seems a suitable model of representation. In this example, the saturation value is chosen to be $S = 10$.

The maximum allowed control signal amplitude is fixed at $u_m = 10$. The initial state vector is

$$\mathbf{y}(0) = \begin{bmatrix} .9 \\ 0 \\ 0 \end{bmatrix}$$

and the desired reference value is $y_R = 1.0$. The program is run for a total of 20 sampling periods; i.e., $N = 20$. The results are summarized in Table 8-1 and in Figs. 8-3 and 8-4.

Table 8-1 [27]

Sampling instant, i	Output, $y_1(i)$	Control signal, u_{i-1}	Error, $e_i = y_1(i) - y_R$
1	.09494	1.1075	−.905063
2	1.22355	2.3787	.223548
3	1.04979	.5408	.049791
4	1.00955	1.3428	.009553
5	.97008	1.2791	−.029922
6	1.01657	1.2225	.016570
7	.96834	1.2601	−.031656
8	1.02376	1.2402	.023765
9	.97652	1.2814	−.023479
10	1.03857	1.2390	.038575
11	.96528	1.2212	−.034724
12	1.01545	1.2390	.015446
13	.97425	1.2528	−.025753
14	1.02848	1.2791	.028779
15	.98529	1.1200	−.014708
16	.99706	1.2555	−.002943
17	.98556	1.2476	−.014439
18	1.00052	1.2287	.000521
19	.97809	1.2304	−.021913
20	.99744	1.2589	−.002564

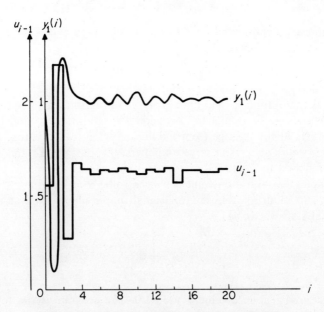

FIG. 8-3. The output and the control signal [27].

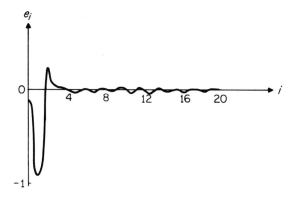

FIG. 8-4. The error [27].

The total computing time is 12.12 sec on the IBM 360/91 System. The value of the optimal performance index is

$$J_0 = .8803$$

8.4 A DIGITALLY CONTROLLED NONLINEAR SYSTEM [20]

The system under consideration is sketched in Fig. 8-5. The controller of the system is part of a digital computer. To be more precise, the digital controller is realized as a subroutine programmed on the computer used for the control purpose. Either a special-purpose control computer or a time-sharing station connected to a central computer system may be used. For

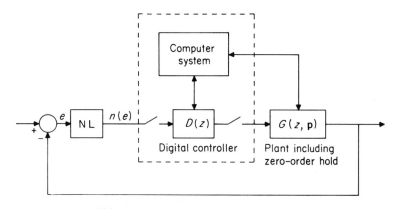

FIG. 8-5. Digitally controlled system [20].

instance, in the case of the digital control of the antenna-tracking system, which is to be discussed later, an SDS Sigma 5 Computer is used, and the digital controller is programmed in assembly language.

The system considered (Fig. 8-5) is a nonlinear sampled-data system. In this case, a different stability criterion than the one used in Sec. 6.4 should be used, namely, the Jury and Lee criterion [21]. The following conditions are imposed on the nonlinearity:

1. $n(e)$ is continuous.
2. $n(0) = 0$.
3. $K > \dfrac{n(e)}{e} > 0$ for $e \neq 0$. (8-23)
4. $\left|\dfrac{dn(e)}{de}\right| < K'$.

If $G^*(z)$ is the overall z-transfer function of the linear subsystem, including the controller, the Jury and Lee theorem states that the system is absolutely stable if a q exists such that

$$JL(z) \equiv \text{Re}\{G^*(z)[1 + q(z-1)]\} + \frac{1}{K} - \frac{K'|q|}{2}|(z-1)G^*(z)|^2 \geqslant 0 \tag{8-24}$$

is satisfied on the unit circle $z = \exp(j\omega T)$, where T is the sampling period.

The computational algorithm in this case would have to run in two steps:

1. Establish for which value of z the left-hand side of inequality in Eq. (8-24) has the minimum value.
2. Substitute the value of z obtained in step 1 into the inequality in Eq. (8-24), and use it as a basis for the NLP problem to establish the optimal parameters that would stabilize the system. Of course, one has to formulate a suitable performance index, and additional constraints may be used.

This algorithm may be appiled to a variety of digital control systems. For instance, it can be used in the stabilization and control of antenna tracking. A flow chart illustrating the proposed algorithm is sketched in Fig. 8-6.

As an illustrative example of the stabilization algorithm proposed, the case of computer control of a 40-ft antenna-tracking system of the NASA Goddard Space Flight Center is chosen. The basic system configuration is sketched in Fig. 8-7. The z-transform transfer function of the linear part of the plant, including a zero-order hold, is

$$G(z) = \frac{.179z^2 + .55z + .155}{z^3 - .114z^2 + .233z - .275} \tag{8-25}$$

The nonlinearity is assumed to satisfy all the conditions specified in

Sec. 8.4   A Digitally Controlled Nonlinear System

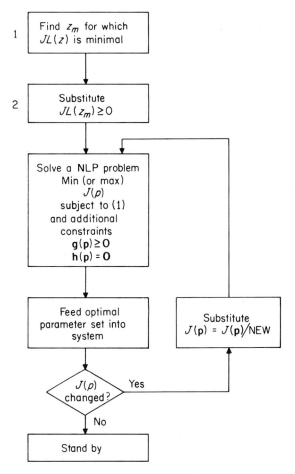

FIG. 8-6. The algorithm for the digitally controlled system [20].

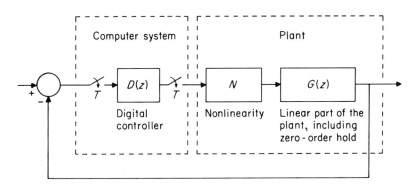

FIG. 8-7. Antenna tracking system configuration [20].

Eq. (8-23). Otherwise, no particular configuration is assumed for the nonlinearity, which makes the solution applicable to a wide class of systems. The digital controller is chosen to be of second order:

$$D(z) = \frac{a_0 z^2 + a_1 z + a_2}{z^2 + b_1 z + b_2} = \frac{a_0 + a_1 z^{-1} + a_2 z^{-2}}{1 + b_1 z^{-1} + b_2 z^{-2}} \qquad (8\text{-}26)$$

The parameters of the digital controller, a_0, a_1, a_2, b_1, and b_2, are unknown variables and should be established in the process of the solution. The digital controller may be programmed on a digital computer according to well-established methods [22]. The choice of a second-order controller is somewhat arbitrary and based rather on experience. There is no loss of generality in this choice, since the same method would apply to any order of $D(z)$. The only difference it would make is in the number of variables. One should, of course, try to solve a problem with a minimum number of unknowns, so one would not want to make the order of $D(z)$ too high. Since the plant is of the third order, it is reasonable to choose the controller of one order less. In any case, the choice of the order of $D(z)$ is a minor issue here. What is important is the method of establishing the actual values of its parameters. These parameters are actually the input data to the subroutine which realizes $D(z)$. The communication between the controller and the continuous-time plant is accomplished through A/D and D/A converters.

The transfer function of the entire linear subsystem, which includes the plant and the controller, is

$$G^*(z) = D(z)G(z)$$
$$= \frac{(a_0 + a_1 z^{-1} + a_2 z^{-2})(.179 z^{-1} + .55 z^{-2} + .115 z^{-3})}{(1 + b_1 z^{-1} + b_2 z^{-2})(1 - .114 z^{-1} + .233 z^{-2} + .275 z^{-3})} \qquad (8\text{-}27)$$

As mentioned previously, the first part of the algorithm involves finding the value of z for which the left-hand side of the inequality in Eq. (8-24) is minimal. In other words, one would like to find the value of z for the worst case when the stability of the system is most "threatened." It should be remembered that the left-hand side of Eq. (8-24), $JL(z)$, must be nonnegative in order to satisfy the stability criterion. According to this criterion, z is restricted to be on the unit circle in the complex z plane; i.e.,

$$z = e^{j\omega T} \qquad (8\text{-}28)$$

or

$$|z| = 1 \qquad (8\text{-}29)$$

One of the simplest ways of finding the value z_m which minimizes $JL(z)$ is by direct search, since only one independent variable z is involved. The restrictions in Eq. (8-28) or (8-29) should, of course, be preserved. This is done simply by performing the search using the real value of z, $\text{Re}(z)$, as an independent variable, and then computing the imaginary value of z by using

$$\text{Im}(z) = \sqrt{1 - \text{Re}^2(z)} \qquad (8\text{-}30)$$

In this example the search is performed by direct scanning over the values of Re(z) at a fixed interval. Of course, one may use more sophisticated search methods if desired. The search takes .3 sec on the IBM 360/91 System. The results obtained are

$$\begin{aligned} z_m &= -.270 + j.963 \\ JL(z_m) &= -7110.0 \end{aligned} \qquad (8\text{-}31)$$

The same problem is also solved as an NLP problem using the SUMT [19] method. The NLP problem is formulated as

$$\min \{JL(z, K, K', q) \,|\, |K| \leqslant K_{\max}; |K'| \leqslant K'_{\max}; |q| \leqslant q_{\max}; |z| = 1\} \qquad (8\text{-}32)$$

Notice that the objective function in this problem depends explicitly on a complex variable z and on a complex function $G(z)$ [see Eq. (8-24)]. In this problem $G(z)$ is used instead of $G^*(z)$ in Eq. (8-24). For the purpose of the numerical solution, instead of z, one works with two variables x_1 and x_2, which are the real and imaginary parts of z, respectively; i.e.,

$$z = x_1 + jx_2 \qquad (8\text{-}33)$$

The complexity of z and its connection with $x = \text{Re}(z)$ is stipulated in the program by the following two FORTRAN IV statements:

COMPLEX Z

EQUIVALENCE (Z, X(1))

This automatically implies that $X(2) \equiv x_2$ is the imaginary part of z.

The conditions in Eq. (8-28) or (8-29) are taken care of by imposing an explicit equality constraint:

$$x_1^2 + x_2^2 - 1 = 0 \qquad (8\text{-}34)$$

Although constraints in this case [see Eqs. (8-32) and (8-34)] are very simple, the objective function $JL(x_1, x_2, K, K', q)$ is a very complicated function of all five variables of the problem. Although JL itself is real, it does depend on complex values z and $G(z)$ explicitly, which makes the numerical solution of the NLP problem quite difficult. In this case, a special technique of unconstrained minimization, developed by Fiacco and McCormick [19], is adopted. This technique does not require the explicit calculation of the derivatives of the functions involved. This property is important in this solution, considering the complexity of the JL function. The solutions obtained in this case are

$$z_m = -.288 + j.958$$
$$JL(z_m) = -7101.0$$
$$K = 48.4 \qquad (8\text{-}35)$$
$$K' = q = 100.0$$

This example takes about 28 sec to run on the same system. One may argue that the SUMT run is unnecessary. However, one would not be able to perform the search without the appropriate values of K, K', and q which are obtained in the SUMT run. The search itself is needed, since the particular algorithm employed in conjunction with SUMT is known for its lack of precision whenever complicated functions are involved. Still, by comparing the results in Eqs. (8-31) and (8-35), one may see that they are not too far apart [.13% for $JL(z_m)$]. Obviously, the result of the search, Eq. (8-31), is picked, since in it $JL(z_m)$ is smaller. The search reveals that the value of $JL(z)$ is quite flat in that region:

$$JL[\text{Re}(z) = -.30] = -7074.2$$
$$JL[\text{Re}(z) = -.25] = -7078.5$$

Therefore, the precision to the second significant figure of $\text{Re}(z)$ is not critical.

The value of z_m in Eq. (8-31) is used in Eqs. (8-34) and (8-27). This time the value of $G^*(z)$ of Eq. (8-27) is substituted into JL in Eq. (8-24).

Now a new NLP problem is formed:

$$\min \{J(K, K') \,|\, JL(K, K', q, a_0, a_1, a_2, b_1, b_2) \geqslant 0; \mathbf{p} \in P\} \qquad (8\text{-}36)$$

where $\mathbf{p} = [K, K', q, a_0, a_1, a_2, b_1, b_2]$ is the parameter vector in this example. P is a closed set; in this case the values of all of the parameters involved are limited in size. The limit imposed on all parameters in this case is

$$|p_i| \leqslant 100.0, \qquad i = 1, \ldots, 8 \qquad (8\text{-}37)$$

The performance criterion in this case is of the form

$$\max J = K + wK' \qquad (8\text{-}38)$$

where w is a weighting factor, in this case chosen to be $w = 1$. The meaning of the problem is that one desires to find a set of parameters $\mathbf{p}$ which permits operating the system with maximum values of K and K' while keeping the system stable.

Since the SUMT program is geared to solve minimization problems, the performance criterion is reformulated:

$$\min J(K, K') = -K - wK' \qquad (8\text{-}39)$$

The results obtained in this run, which takes 5 sec on the IBM 360/91 System, are

$K = K' = 100.0$ (i.e., working at the limit)
$q = 15.0$
$a_0 = .131$
$a_1 = .063 \quad b_1 = .100$
$a_2 = .129 \quad b_2 = .094$

Or, in other words, the digital controller is

$$D(z) = \frac{.131 + .063z^{-1} + .129z^{-2}}{1 + .100z^{-1} + .094z^{-2}} \tag{8-40}$$

The actual value of JL for the mentioned parameters is .08. This is quite close to the stability limit of zero. Therefore, in actual operation somewhat lower values of K and/or K' should be used. In solving problems such as this, a limit of K and K' should be set about 20% higher than the desired value. It should be stressed that the search for z_m is done infrequently and only if there is a considerable change in the plant's parameters. In view of the short computing times for the parameters of the digital controller, this algorithm can definitely be employed in real-time control processes where changes in the system requirements do not occur more often than about every 10 sec. Thus the algorithm can indeed cover a very wide class of computer-controlled processes.

8.5 DESIGN OF A DIGITAL CONTROLLER FOR A NONLINEAR SAMPLED-DATA SYSTEM [17, 23]

The work described here extends the work described in Sec. 7.3 to nonlinear sampled-data systems. The approach is basically that of writing sequential equations defining the relationships among the variables of the system at each sampling instant. These equations are usually nonlinear, not only because of nonlinearities present in the system, but also, as it will be shown, because of cross products between variables at different sampling instants. A performance criterion is then imposed by writing an objective function of the system errors at different sampling instants. An NLP problem is thus defined.

The basic configuration of a uniformly sampled-data system under consideration is shown in Fig. 8-8. In this system,

$D(z)$ = Pulse transfer function of the plant
$G(z)$ = Pulse transfer function of the plant, including the hold circuit (zero-order hold)
N = Nonlinearity

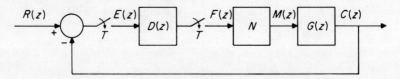

FIG. 8-8. A nonlinear sampled-data control system [23].

Usually, the zero-order hold precedes the nonlinearity in the forward path. However, interchanging the position of the zero-order hold with that of N does not alter the system mathematically, but is does considerably simplify the design equations.

The system equations in z transformation are

$$\begin{aligned} E(z) &= R(z) - C(z) \\ C(z) &= G(z)M(z) \\ F(z) &= D(z)E(z) \\ M(z) &= N[F(z)] \end{aligned} \quad (8\text{-}41)$$

where N is the input-output relationship for the nonlinearity.

Usually $G(z)$ is the quotient of two polynomials in z^{-1}:

$$G(z) = \frac{P(z)}{Q(z)} \quad (8\text{-}42)$$

where

$$\begin{aligned} P(z) &= p_k z^{-k} + p_{k+1} z^{-(k+1)} + \cdots + p_m z^{-m} \\ Q(z) &= q_0 + q_1 z^{-1} + \cdots + q_z^{-l} \end{aligned} \quad (8\text{-}43)$$

and k is the delay of the plant in sampling periods [22].

The pulse transfer function of the digital controller is also a quotient of polynomials in z^{-1} [22]:

$$D(z) = \frac{A(z)}{B(z)} \quad (8\text{-}44)$$

where

$$\begin{aligned} A(z) &= a_0 + a_1 z^{-1} + \cdots + a_u z^{-u} \\ B(z) &= 1 + b_1 z^{-1} + \cdots + b_v z^{-v} \end{aligned} \quad (8\text{-}45)$$

Substituting Eqs. (8-42) and (8-44) into Eq. (8-41), we get

$$R(z)Q(z) = E(z)Q(z) + P(z)M(z) \quad (8\text{-}46)$$
$$A(z)E(z) - B(z)F(z) = 0 \quad (8\text{-}47)$$

Let the input and the other signals in the system be described by the sequences

Sec. 8.5 Digital Controller for a Nonlinear Sampled-Data System 189

$$R(z) = \sum_{i=0}^{n} r_i z^{-i}$$

$$E(z) = \sum_{i=0}^{n} e_i z^{-i}$$

$$F(z) = \sum_{i=0}^{n} f_i z^{-i} \qquad (8\text{-}48)$$

$$M(z) = \sum_{i=0}^{n} m_i z^{-i}$$

where $r_i = [r(t)]_{t=iT}$, T being the sampling period (and also for e_i, f_i, and m_i), and n is the total number of sampling instants considered.

Using Eqs. (8-42), (8-45), and Eq. (8-48) in Eqs. (8-46), (8-47), and the last equation of Eq. (8-41), expanding the products and equating the coefficients of equal powers in z^{-1}, we obtain the following set of equations:

$$\begin{aligned}
r_0 q_0 &= e_0 q_0 \\
r_1 q_0 + r_0 q_1 &= e_1 q_0 + e_0 q_1 \\
r_k q_0 + \cdots + r_0 q_k &= e_k q_0 + \cdots + e_0 q_k + m_0 p_k \\
r_l q_0 + \cdots + r_0 q_l &= e_l q_0 + \cdots + e_0 q_l + m_{l-k} p_k + \cdots + m_0 p_l \\
r_m q_0 + \cdots + r_{m-l} q_l &= e_m q_0 + \cdots + e_{m-l} q_l + m_{m-k} p_k + \cdots + m_0 p_m \\
r_n q_0 + \cdots + r_{n-l} q_l &= e_n q_0 + \cdots + e_{n-l} q_l + m_{n-k} p_k + \cdots + m_{n-m} p_m
\end{aligned} \qquad (8\text{-}49)$$

Similarly, substituting Eq. (8-45) into Eq. (8-47), we have

$$\begin{aligned}
e_0 a_0 - f_0 &= 0 \\
e_1 a_0 + e_0 a_1 - f_1 - f_0 b_1 &= 0 \\
e_u a_0 + \cdots + e_0 a_u - f_u - \cdots - f_0 b_u &= 0 \\
e_n a_0 + \cdots + e_{n-u} a_u - f_n - \cdots - f_{n-u} b_u &= 0
\end{aligned} \qquad (8\text{-}50)$$

where it is assumed that $u = v$.

$$\begin{aligned}
m_0 &= N(f_0) \\
m_n &= N(f_n)
\end{aligned} \qquad (8\text{-}51)$$

For simplicity, it has been assumed in expanding the products of Eq. (8-46) that $k \leq l \leq m$, which is usually satisfied by the pulse transfer function of the plant. Note that whereas the equations in Eq. (8-49) are linear in the variables e_i and m_i, the equations in Eq. (8-50) are nonlinear because of the cross products arising from Eq. (8-47). Thus, even if the system considered does not have the nonlinearity, the design equations of Eqs. (8-49) and (8-50), with $m_i = f_i$ ($i = 0, 1, \ldots, n$), would still be nonlinear. This is due only to the approach followed, which suits particularly nonlinear systems. For linear systems, with a slightly different approach, the system may be designed more efficiently by linear or quadratic programming (see Sec. 7.3).

The systems of equations in Eqs. (8-49) through (8-51) constitute the equality constraints of an NLP problem. Additional inequality constraints may be imposed on the variables to meet particular requirements, such as the conditions that the sampled errors be limited in absolute value.

An optimality criterion based on a chosen performance index is to be written to define the NLP problem. A weighted, summed, squared-error criterion is chosen:

$$\text{minimize } J = \sum_{i=0}^{n} c_i e_i^2 \qquad (8\text{-}52)$$

where c_i is the weight assigned to the ith sampled error.

The same method can be easily extended to systems with more than one nonlinearity, such as the system shown in Fig. 8-9.

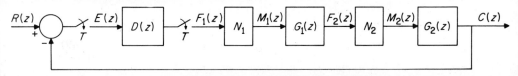

FIG. 8-9. A system with two nonlinearities [23].

Similar to Eq. (8-41), the system equations are

$$\begin{aligned}
E(z) &= R(z) - C(z) \\
C(z) &= G_2(z) M_2(z) \\
M_2(z) &= N_2[F_2(z)] \\
F_2(z) &= G_1(z) M_1(z) \\
M_1(z) &= N_1[F_1(z)] \\
F_1(z) &= D(z) E(z)
\end{aligned} \qquad (8\text{-}53)$$

Following the same steps as described before, design equations completely similar in their structure to Eqs. (8-49) and (8-51) are obtained.

To illustrate the design method proposed, a system with a saturation nonlinearity is chosen. In many practical cases, the nonlinearity is smooth rather than piecewise linear. Therefore, the saturation nonlinearity is simulated by a hyperbolic tangent function; i.e., Eq. (8-51) is written as

$$m_i = S \tanh \frac{f_i}{S}, \qquad i = 0, 1, \ldots, n \qquad (8\text{-}54)$$

where S is the saturation value of m_i. For $f_i = 0$, the slope of this curve is unity.

The plant considered has the transfer function

$$G(s) = \frac{.0018(1 - e^{-Ts})}{s^2(1 + 200s)} \qquad (8\text{-}55)$$

Sec. 8.5 Digital Controller for a Nonlinear Sampled-Data System

which includes a zero-order hold, a pure integration, and a single time constant. With a sampling period of $T = 50$ sec, the plant z-transform transfer function is

$$G(z) = \frac{.014z^{-1}(1 + .93z^{-1})}{(1 - z^{-1})(1 - .78z^{-1})} \tag{8-56}$$

The following additional data have been chosen:

Saturation limit: $S = 30$ V
System input is a step function: $r(t) = 10u(t)$
Number of sampling instants considered: $n = 20$
Form of the digital controller:

$$D(z) = \frac{a_0 + a_1 z^{-1} + a_2 z^{-2}}{1 + b_1 z^{-1} + b_2 z^{-2}} \tag{8-57}$$

Performing the appropriate substitution in Eqs. (8-49) through (8-51), we derive the design equations:

$$
\begin{aligned}
e_0 &= 10 \\
e_1 &+ .0104m_0 = 10 \\
e_2 &- 1.78e_1 + .0104m_1 + .00967m_0 = -7.8 \\
e_3 &- 1.78e_2 + .78e_1 + .0104m_2 + .00967m_1 = 0 \\
&\vdots \\
e_{20} &- 1.78e_{19} + .78e_{18} + .0104m_{19} + .00967m_{18} = 0 \\
10a_0 &\quad - f_0 = 0 \\
(e_1 - 10b_1)a_0 &+ 10a_1 - f_1 = 0 \\
(e_2 - 10b_2)a_0 &+ e_1 a_1 + 10a_2 - f_2 - f_1 b_1 = 0 \\
e_3 a_0 &+ e_2 a_1 + e_1 a_2 - f_3 - f_2 b_1 - f_1 b_2 = 0 \\
e_{20} a_0 &+ e_{19} a_1 + e_{18} a_2 - f_{20} - f_{19} b_1 - f_{18} b_2 = 0
\end{aligned}
\tag{8-58, 8-59}
$$

$$m_i = 30 \tanh \frac{f_i}{30}, \quad i = 0, 1, \ldots, 20 \tag{8-60}$$

The performance index actually used is

$$\text{Minimize } J = \sum_{i=1}^{20} 2^i e_i^2 \tag{8-61}$$

The SUMT [19] program is run for this example, minimizing Eq. (8-61) subject to the equality constraints of Eqs. (8-58) through (8-60). The digital controller obtained is

$$D(z) = \frac{170.87 + 4.63z^{-1} + 5.26z^{-2}}{1 + .439z^{-1} + .131z^{-2}}$$

The values of the signals in the system at the sampling instants $i = 1, \ldots, 20$ are tabulated as follows:

i	e_i	f_i	m_i
0	10.000	1708.70	30.00
1	3.716	−69.18	−29.94
2	.338	−66.16	−29.94
3	−.106	42.43	26.68
4	.013	−4.56	−4.52
5	.021	1.06	1.05
6	.017	4.10	4.13
7	.013	1.32	1.32
8	9.303×10^{-3}	1.17	1.17
9	7.769×10^{-3}	1.19	1.19
10	9.691×10^{-3}	1.43	1.43
11	7.874×10^{-3}	1.16	1.16
12	7.447×10^{-3}	1.22	1.22
13	6.971×10^{-3}	1.25	1.25
14	5.656×10^{-3}	1.11	1.11
15	4.169×10^{-3}	1.06	1.06
16	2.720×10^{-3}	1.00	1.00
17	1.572×10^{-3}	.92	.92
18	8.527×10^{-4}	.90	.90
19	4.301×10^{-4}	.87	.87
20	2.056×10^{-4}	.80	.80

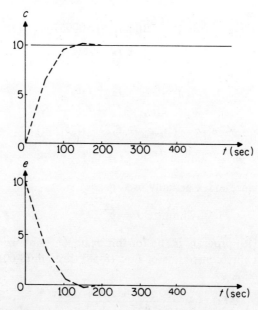

FIG. 8-10. Input, sampled output, and sampled error for the example [23].

The optimal performance index is found to be $J_0 = 31.34$. The time variation of the output and error functions is shown in Fig. 8-10.

8.6 AN ITERATIVE SOLUTION

A solution of an optimal-control problem for a class of nonlinear discrete-time systems has been proposed by Rosen [24]. The algorithm consists of a sequence of solutions of MP problems with linear constraints. Rosen proved that the iterative procedure proposed converges [24].

The problem treated is the following: Given a nonlinear discrete-time system with the state difference dynamic equations

$$\mathbf{x}(i+1) - \mathbf{x}(i) = \mathbf{f}[\mathbf{x}(i), \mathbf{u}(i)], \quad i = 0, 1, \ldots, N-1 \quad (8\text{-}62)$$

find an optimal sequence of the control vectors $\mathbf{u}(i)$ so that the following performance index is minimized:

$$J = \sum_{i=1}^{N-1} F[\mathbf{x}(i), \mathbf{u}(i)] \quad (8\text{-}63)$$

where $\mathbf{x}(i) = n$-dimensional state vector at $t = t_i$
$\mathbf{u}(i) = r$-dimensional control vector at $t = t_i$
$\mathbf{f} = n$-dimensional nonlinear function
$F =$ Scalar nonlinear function

In addition, it is assumed that the following control and state constraints are imposed:

$$\mathbf{u}(i) \in U_i, \quad i = 0, 1, \ldots, N-1 \quad (8\text{-}64)$$
$$\mathbf{x}(i) \in X_i, \quad i = 0, 1, \ldots, N \quad (8\text{-}65)$$

where U_i and X_i are assumed to be compact and convex sets. It is further assumed that F is a convex function from each direct product $U_i \times X_i$ to E^1 and that $\mathbf{f}$ is a function from each $X_i \times U_i$ to E^n. To assure convergence of the proposed iterative procedure, Rosen assumed that the components of $\mathbf{f}$ should be either convex or concave on $X_i \times U_i$ [24].

The following new vector is introduced; it includes all of the unknown variables of the problem:

$$\mathbf{z} = [\mathbf{x}(0), \mathbf{x}(1), \ldots, \mathbf{x}(N), \mathbf{u}(0), \mathbf{u}(1), \ldots, \mathbf{u}(N-1)]^T \quad (8\text{-}66)$$

It is apparent that $\mathbf{z}$ is a $[n(N+1) + rN]$-dimensional vector. Let $M = [n(N+1) + rN]$.

A new constraint set Z for the $\mathbf{z}$ vector is introduced:

$$\mathbf{z} \in Z = \prod_{i=0}^{N} X_i \times \prod_{i=0}^{N-1} U_i \quad (8\text{-}67)$$

Since the sets X_i and U_i are convex and compact for all i, Z is also convex and compact. After substituting the components of $\mathbf{z}$ into the performance index in Eq. (8-63), J is now a function of $\mathbf{z}$: $J(\mathbf{z})$.

The equality constraints generated by the state difference equations in Eq. (8-62) are replaced as follows:

$$v_{i,j} = f_j[\mathbf{x}(i), \mathbf{u}(i)] + x_j(i) - x_j(i+1) = 0, \quad \begin{array}{l} i = 0, 1, \ldots, N-1 \\ j = 1, \ldots, n \end{array} \quad (8\text{-}68)$$

The new entities $v_{i,j}$ comprise the vector $\mathbf{v}(\mathbf{z})$, so that the constraints in Eq. (8-68) are expressed as

$$\mathbf{v}(\mathbf{z}) = \mathbf{0} \qquad (8\text{-}69)$$

$\mathbf{v}$ is obviously an Nn-dimensional vector. Let $L = Nn$.

The optimal-control problem formulated at the beginning of this section can now be formulated as the following MP problem:

$$\min_{\mathbf{z}} \{J(\mathbf{z}) \mid \mathbf{z} \in Z; \mathbf{v}(\mathbf{z}) = \mathbf{0}\} \qquad (8\text{-}70)$$

This formulation is basically identical to the one given in Eq. (8-7). So far, the original problem is fully represented in Eq. (8-70) and no approximation has been made. The general and most obvious approach in this case is direct application of an NLP algorithm to the problem in Eq. (8-70). In this case, however, a global minimum is guaranteed only if the feasible region is convex. It was stated before that Z is convex, so every component of $\mathbf{v}(\mathbf{z})$ is also convex. The feasible region defined in Eq. (8-70) is convex only if $\mathbf{v}(\mathbf{z})$ is linear in $\mathbf{z}$. To achieve convexity, Rosen linearized the system's state equations through a first-order Taylor expansion around any $\mathbf{z}_1$:

$$\mathbf{w}(\mathbf{z}, \mathbf{z}_1) = \mathbf{v}(\mathbf{z}_1) + \left\| \frac{\partial \mathbf{v}}{\partial \mathbf{z}} \right\|_{\mathbf{z}_1} (\mathbf{z} - \mathbf{z}_1) \qquad (8\text{-}71)$$

where

$$\left\| \frac{\partial \mathbf{v}}{\partial \mathbf{z}} \right\|_{\mathbf{z}_1}$$

is an $L \times M$ Jacobian matrix of $\mathbf{v}(\mathbf{z})$ evaluated at $\mathbf{z} = \mathbf{z}_1$.

For the computational realization of the proposed iterative algorithm, the following sets are defined:

$$S = \{\mathbf{z} \mid \mathbf{z} \in Z, \mathbf{v}(\mathbf{z}) \geqslant 0\} \qquad (8\text{-}72)$$

$$W(\mathbf{z}_1) = \{\mathbf{z} \mid \mathbf{w}(\mathbf{z}, \mathbf{z}_1) \geqslant 0\} \qquad (8\text{-}73)$$

Now define the following point-to-set mapping:

$$\Gamma: Z \to Z$$

as an intersection between $W(\mathbf{z}_1)$ and Z:

$$\Gamma \mathbf{z}_1 = W(\mathbf{z}_1) \cap Z \qquad (8\text{-}74)$$

Rosen has shown that the set $\Gamma \mathbf{z}_1$ is compact and convex [24].

The iterative procedure starts with an arbitrary initial point $z^0 \in S$ and the following MP problem is solved:

$$J(z^1) = \min_{z \in \Gamma z^0} J(z) \tag{8-75}$$

It will be assumed that J contains a penalty term which compensates for the enlargement of the feasible region. This iterative procedure is then carried on generating a sequence of suboptimal points $z^1, z^2, \ldots$ according to

$$J(z^{j+1}) = \min_{z \in \Gamma z^j} J(z), j = 0, 1, \ldots \tag{8-76}$$

The convexity of $J(z)$ and Γz^j assures that a global minimum of $J(z)$ for $z \in \Gamma z^j$ is attained at $z = z^{j+1}$. Rosen has proved the convergence of this procedure in reference 24. The method has been further extended and generalized by Meyer [25, 26].

REFERENCES

1. L. S. Pontryagin, V. G. Boltianskii, R. V. Gamkrelidze, and E. F. Mischenko, *The Mathematical Theory of Optimal Processes*, Wiley-Interscience, New York, 1962.
2. L. I. Rozonoer, "L. S. Pontryagin's Maximum Principle in the Theory of Optimum Systems," *Automation Remote Control*, **20**, Part I, pp. 1288–1302; Part II, pp. 1405–1421; Part III, pp. 1517–1532, 1959.
3. S. S. L. Chang, "Digitized Maximum Principle," *IRE Proc.*, **48**, pp. 2030–2031, 1960.
4. S. Katz, "A Discrete Version of Pontryagin's Maximum Principle," *J. Electronics Control*, **13**, pp. 179–184, 1962.
5. L. T. Fan and C. S. Wang, *The Discrete Maximum Principle*, Wiley, New York, 1964.
6. H. Halkin, B. W. Jordan, E. Polak, and J. B. Rosen, "Theory of Optimum Discrete Time Systems," 3rd IFAC Conference, London, 1966.
7. J. D. Pearson, "The Discrete Maximum Principle," *Intern. J. Control*, **2**, pp. 117–124, 1965.
8. J. M. Holtzman, "Convexity and the Maximum Principle for Discrete Systems," *IEEE Trans. Automatic Control*, **AC-11**, pp. 30–35, 1966.
9. A. I. Propoi, "A Problem of Optimal Discrete Control," *Soviet Phys. Doklady*, **9**, pp. 1040–1042, 1965.
10. J. T. Tou, *Optimum Design of Digital Control Systems*, Academic Press, New York, 1963.
11. A. J. Brockstein and B. C. Kuo, "Optimum Control of Multivariable Sampled Data Systems with Adaptive Sampling," 1967 JACC, pp. 366–372, Philadelphia, Pa.

12. J. B. Rosen, "Optimal Control and Convex Programming," *MRC Technical Report* 547, University of Wisconsin, Madison, Feb. 1965.
13. M. W. Kuhn and A. W. Tucker, "Nonlinear Programming," in *Proceedings of the Second Berkeley Symposium on Mathematical Statistics and Probability*, University of California Press, Berkeley, 1951, pp. 481–492.
14. J. B. Pearson and R. Sridhar, "A Discrete Optimal Control Problem" *IEEE Trans. Automatic Control*, **AC-11**, pp. 171–174, 1966.
15. O. L. Mangasarian and S. Fromovitz, "A Maximum Principle in Mathematical Programming," in *Mathematical Theory of Control* (A. V. Balakrishnan, and L. W. Neustadt, eds.), Academic Press, New York, 1967. pp. 85–95.
16. M. D. Canon, C. D. Cullum, and E. Polak, "Constrained Minimization Problems in Finite Dimensional Spaces," *J. SIAM Control*, **4**, pp. 528–547, 1966.
17. D. Tabak, "Application of Mathematical Programming in the Design of Optimal Control Systems," Ph.D. Thesis, University of Illinois, Urbana, 1967.
18. D. Tabak and B. C. Kuo, "Application of Mathematical Programming in the Design of Optimal Control Systems," *Intern. J. Control*, **10**, pp. 545–552, 1969.
19. A. V. Fiacco and G. P. McCormick, *Nonlinear Programming: Sequential Unconstrained Minimization Techniques*, Wiley, New York, 1968.
20. D. Tabak, "An Algorithm for Nonlinear Process Stabilization and Control," *IEEE Trans. Computers*, **C-19**, pp. 487–492, 1970.
21. E. I. Jury and B. W. Lee, "On the Stability of a Certain Class of Nonlinear Sampled-Data Systems," *IEEE Trans. Automatic Control*, **AC-9**, pp. 51–61, Jan. 1964.
22. B. C. Kuo, "Analysis and Synthesis of Sampled Data Control Systems," Prentice-Hall, Englewood Cliffs, N.J., 1963.
23. G. Porcelli, D. Tabak, and K. A. Fegley, "Design of Digital Controllers for Nonlinear Systems by Mathematical Programming," **1968** *WESCON*, Session 14, Los Angeles, Cal., Aug. 1968.
24. J. B. Rosen, "Iterative Solution of Nonlinear Optimal Control Problems," *J. SIAM Control*, **4**, pp. 223–244, 1966.
25. R. R. Meyer, "The Solution of Non-Convex Optimization Problems by Iterative Convex Programming," Ph.D. Thesis, University of Wisconsin, Madison, 1968.
26. R. R. Meyer, "The Validity of a Family of Optimization Methods," *J. SIAM Control*, **8**, pp. 41–54, 1970.
27. D. Tabak, "Optimal Control of Nonlinear Discrete-Time Systems by Mathematical Programming," *J. Franklin Inst.*, **289**, pp. 111–119, 1970.

9

STOCHASTIC SYSTEMS

9.1 INTRODUCTION

Relatively little work has been done in applying MP techniques to stochastic control systems. There exists, though, a vast amount of literature dedicated to the problems of optimal control of stochastic systems; only a small sample is referenced here [1–7].

In this chapter, several projects where the application of MP techniques to optimal control of stochastic systems was attempted are discussed. In Sec. 9.2, a method for a combined utilization of MP algorithms with a Monte Carlo approach is presented. Design of a digital controller for a sampled-data system whose input is corrupted by noise is discussed in Sec. 9.3.

Section 9.4 outlines a method for applying MP in an optimal-estimation problem. We can argue, of course, that there exist already very efficient methods for the solution of the optimal-estimation problem [1–7]. This is true, of course, for *unconstrained* estimation problems. In cases where there are inequality constraints imposed on the state and control variables, we would experience a very difficult computational problem while trying to apply the conventional methods [1–7]. On the other hand, utilizing MP, the computational solution of constrained optimal-estimation problems becomes considerably easier. Section 9.5 concludes with an application of MP in a problem of identification of sampled-data systems.

9.2 COMBINED MP-MONTE CARLO APPROACH [8, 9]

A typical stochastic control system can be represented schematically as shown in Fig. 9-1 [1].

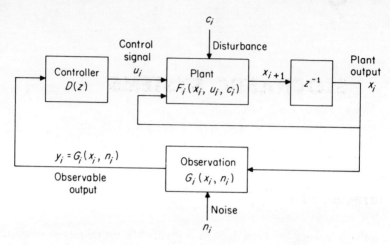

FIG. 9-1. A stochastic control system [1, 9].

The system under consideration is of the discrete-time type and is presented using the z transform. The plant is governed by a set of nonlinear difference equations:

$$x_{i+1} = F_i(x_i, u_i, c_i), \quad i = 0, 1, \ldots, N-1 \tag{9-1}$$

where $x_i =$ Plant output at the ith time instant
$u_i =$ Control signal input to the plant at the ith instant
$c_i =$ Parameter disturbance input to the plant at the ith instant
$F_i =$ General nonlinear function

For the sake of simplicity, a single-dimension system is considered. However, in general, x_i, u_i, and c_i may be vector variables. The actually observed system output y_i may be expressed by another, generally nonlinear function G_i:

$$y_i = G_i(x_i, n_i), \quad i = 1, \ldots, N \tag{9-2}$$

where n_i is the noise at instant t_i.

The observable signal y_i is actually fed into the controller, which is to be designed. Using the z transform, one may assume the following general form for the controller:

$$D(z) = \frac{a_0 + a_1 z^{-1} + \cdots + a_p z^{-p}}{1 + b_1 z^{-1} + \cdots + b_p z^{-p}} = \frac{U(z)}{Y(z)} \tag{9-3}$$

where p is the assumed order of the controller and $a_0 \cdots a_p$ and $b_1 \cdots b_p$ are the controller parameters to be evaluated. Knowing those parameters would permit physical realization of the controller [10].

From Eq. (9-3) the control signal at the ith time instant may be expressed as a function of the present y_i and the y's and u's for the past p sampling intervals [10]. Therefore,

$$u_1 = a_0 y_i + a_1 y_{i-1} + \cdots + a_p y_{i-p} - b_1 u_{i-1} - \cdots - b_p u_{i-p} \quad (9\text{-}4)$$

Of course, it is assumed that

$$y_i = u_i = 0$$

for $i < p$.

Owing to physical considerations and limitations concerning possible plant inputs, we can pose constraints on the u_i's as

$$|u_i| \leqslant u_{\max}, \quad i = 0, 1, \ldots, N-1 \quad (9\text{-}5)$$

When a system is being designed, we should, of course, pose certain design criteria. For instance, we can formulate the design problem in the following way:

Find the digital controller parameters $a_0, \ldots, a_p$ and $b_1, \ldots, b_p$ so that the performance index

$$J = \sum_{i=1}^{N} W_i(x_i, u_{i-1}) \quad (9\text{-}6)$$

is minimized, subject to the constraints in Eqs. (9-1), (9-2), (9-4), and (9-5).

Inspecting these four equations, we can see that they form a NP problem (Chapter 2). There is, however, a major difference. While the equations listed in Chapter 2 contain deterministic variables and constraints, the equations in this section contain, in addition, a set of random variables, c_i ($i = 0, 1, \ldots, N-1$) and n_i ($i = 1, \ldots, N$).

To apply a NLP algorithm, we would have to substitute some definite values in Eqs. (9-1) and (9-2) instead of c_i and n_i.

If the statistics and the bounds of c_i and n_i are known, we can use a pseudo-random generator, which generates sequences of definite values, $c_0, c_1, \ldots, c_{N-1}$ and $n_1, n_2, \ldots, n_N$. These sequences can be substituted into Eqs. (9-1) and (9-2), and then we have to solve the NLP problem. It is, of course, desirable to use the highest number (N) of signals possible.

For a problem of this kind, the Monte Carlo approach can be used. After solving the first NLP problem, the next two sequences of N pseudo-random numbers, c_i and n_i, are generated, and a second NLP problem is solved. This procedure is repeated M times. At the end, the average values of the controller parameters, a_i and b_i, are adopted as a final solution. If M statistical experiments are performed, the relative error for the final values is $\sqrt{M}$. For instance, if $M = 100$, the relative error is 10%; for $M = 1000$, it is about 3%. Unless a very fast NLP algorithm is available, this procedure might prove quite time consuming. However, with the recent development of multiprocessing computers, we may be able to solve several NLP problems simultaneously, thus making the proposed method more easily achievable. In cases where the statistics of c_i and n_i are unknown, the task of generating sequences of c_i, n_i is much more difficult. We would have to use any pieces of information available about the physical nature of these disturbances and try to simulate them as close to reality as possible. In such a case, a larger number of experiments would be required, because of the higher uncertainty of the c_i, n_i sequences.

9.3 STATISTICAL DESIGN OF SAMPLED-DATA CONTROL SYSTEMS [11]

A linear sampled-data system whose input signal is corrupted by noise is considered in this section. The system configuration is shown in Fig. 9-2.

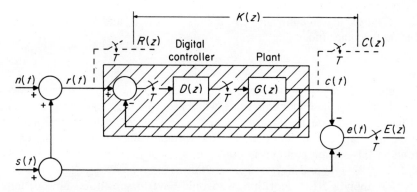

FIG. 9-2. The block diagram with the system to be designed shown in the shaded area [11].

The reference input signal is $s(t)$; the noise signal is $n(t)$. The actual input signal to the system is

$$r(t) = s(t) + n(t) \tag{9-7}$$

The output of the system is $c(t)$. The objective of the control action is to make the output $c(t)$ follow the reference input $s(t)$. The control action is contaminated by the noise $n(t)$. It is the purpose of the present design to provide a digital controller $D(z)$ so that the error

$$e(t) = s(t) - c(t) \tag{9-8}$$

is made as small as possible. As a matter of fact, a quadratic error performance index is used in the following.

The closed-loop transfer function of the control system is

$$K(z) = \frac{D(z)G(z)}{1 + D(z)G(z)} = \sum_{i=1}^{n} K_i z^{-i} \tag{9-9}$$

using the same notation as in Chapter 7. Once $K(z)$ is known, the digital controller may be readily found from

$$D(z) = \frac{K(z)}{G(z)[1 - K(z)]} \tag{9-10}$$

since the plant $G(z)$ is assumed known. This approach may not be applicable in cases where $G(z)$ has zeros outside the unit circle in the z plane. In practice, exact cancellation of these zeros is very difficult to achieve, and since they will appear as poles in $D(z)$, it could bring the system into instability. One should therefore be careful in using the approach presented in this section, as well as any other approach based on Eq. (9-10) (see Chapter 7), in practical applications.

It is further assumed that the correlation functions of the signal and noise are known. The mean-square sampled error corresponding to a random input may be expressed as a contour integral in the complex z plane for systems which operate on the infinite-past or as a quadratic function of the K_i's for finite-memory systems [12]. The resulting form of the performance index is

$$\text{Minimize } J = e^2(nT) = \mathbf{K}A\mathbf{K}^T \tag{9-11}$$

where

$$\mathbf{K} = \begin{bmatrix} 1 \\ K_1 \\ K_2 \\ \cdot \\ \cdot \\ \cdot \\ K_m \end{bmatrix} \quad A = \begin{bmatrix} a_{00} & a_{10} & a_{20} & a_{30} & \cdots & a_{n0} \\ a_{10} & a_{11} & a_{21} & a_{31} & \cdots & a_{n1} \\ a_{20} & a_{21} & a_{11} & a_{21} & \cdots & a_{(n-1)1} \\ \cdot & \cdot & \cdot & \cdot & & \cdot \\ \cdot & \cdot & \cdot & \cdot & & \cdot \\ \cdot & \cdot & \cdot & \cdot & & \cdot \\ a_{n0} & a_{n1} & a_{(n-1)1} & a_{(n-2)1} & \cdots & a_{11} \end{bmatrix} \tag{9-12}$$

A is a symmetrical matrix whose elements depend on the sampled values of the correlation functions $R_{ss}(\tau)$ and $R_{ns}(\tau)$ of the reference signal $s(t)$ and noise $n(t)$:

$$\begin{aligned} a_{00} &= R_{ss}(0) \\ a_{i0} &= -R_{ss}(iT) - R_{ns}(iT), \quad i = 1, 2, \ldots, n \\ a_{11} &= R_{ss}(0) + R_{nn}(0) \\ a_{i1} &= -a_{(i-1),0}, \quad i = 1, 2, \ldots, n \end{aligned} \tag{9-13}$$

The coefficients K_i are the variables of the problem. They usually have both positive and negative values. Since the usual LP or QP algorithms require the variables to be nonnegative, the following transformation of variables is made:

$$K_i = K_{ia} - K_{ib} \tag{9-14}$$

where K_{ia} and K_{ib} are nonnegative quantities. Making this substitution in Eq. (9-11), one obtains

$$\overline{e^2(nT)} = \hat{K}\hat{A}\hat{K}^T \tag{9-15}$$

where

$$\hat{A} = \begin{bmatrix} a_{n0} & a_{10} & \cdots & a_{n0} & -a_{10} & \cdots & -a_{n0} \\ a_{10} & & & & & & \\ \vdots & & A_{00} & & & -A_{00} & \\ a_{n0} & & & & & & \\ -a_{10} & & & & & & \\ \vdots & & -A_{00} & & & A_{00} & \\ -a_{n0} & & & & & & \end{bmatrix} \tag{9-16}$$

$$\hat{K} = [1, K_{1a}, \ldots, K_{na}, K_{1b}, K_{2b}, \ldots, K_{nb}] \tag{9-17}$$

A_{00} is the minor formed by deleting the first row and first column of A in Eq. (9-12).

The minimization of the mean-square sampled error $\overline{e^2(nT)}$ as expressed by Eq. (9-15) is a suitable objective function for the statistical design of a sampled-data control system.

In general, the sources of system constraints are physical realizability conditions and specifications on system response to impulse, step, and/or ramp inputs. Only linear constraint equations are permitted in quadratic programming. For physical realizability, if the plant has a delay of m sampling periods, the overall pulse transfer function $K(z)$ must have at least the same delay. In general, the fixed portion of the system can be expressed as the ratio of two polynomials

$$G(z) = \frac{z^{-m}(h_0 + h_1 z^{-1} + \cdots + h_b z^{-b})}{1 + g_1 z^{-1} + \cdots + g_a z^{-a}}, \quad a \geqslant b \tag{9-18}$$

Since the factor z^{-m} in Eq. (9-18) represents a delay of m sampling periods, it is necessary for physical realizability that

$$K_i = 0, \quad i = 1, 2, \ldots, m - 1 \tag{9-19}$$

Assume that $K(z)$ has the series form given in Eq. (9-9). The K_i's represent the value of the closed-loop response at $t = iT$ owing to a unit impulse input at $z = 0$. If the designer wishes to confine the unit impulse response

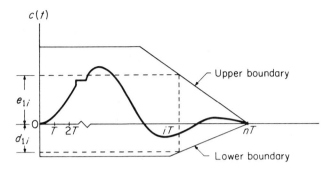

FIG. 9-3. Boundaries on the impulsive response of the system [11].

to the envelope shown in Fig. 9-3, the constraint inequalities are

$$K_i \leqslant e_{1i}, \quad i = 1, 2, \ldots, (n-1)$$
$$K_i \geqslant d_{1i}, \quad i = 1, 2, \ldots, (n-1) \quad (9\text{-}20)$$
$$K_i = 0, \quad i = n, n+1, \ldots$$

Hence if it is desired to terminate the response to an impulse at the nth sampling instant, then A, defined by Eq. (9-12), need be only an $(n \times n)$-dimensional matrix. If $K(z)$ is a finite polynomial, $1 - K(z)$ must contain the factor $(1 - z^{-1})$ for a zero steady-state sampled error in response to a unit step input. Hence $K(z) = 1$ at $z = 1$. Substituting $z = 1$ into Eq. (9-9), where the upper limit now is n, the constraint equation is

$$\sum_{i=1}^{n} K_i = 1, \quad K_i = 0, \quad i > n \quad (9\text{-}21)$$

In this case A need be only an $[(n+1)(n+1)]$-dimensional matrix. The step response for $t = iT$ for $i < n$ may be restricted by the constraint inequalities

$$K_1 + K_2 + \cdots + K_i \leqslant (1 + e_{si}), \quad i = 1, 2, \ldots, (n-1)$$
$$K_1 + K_2 + \cdots + K_i \geqslant (1 - d_{si}), \quad i = 1, 2, \ldots, (n-1) \quad (9\text{-}22)$$

where, as shown in Fig. 9-4, e_{si} and d_{si} fix the range of values which the step response may have at the ith sampling instant. If the system is to follow a unit ramp input within specified limits, the following inequality constraints apply:

$$(i-1)K_1 + (i-2)K_2 + \cdots + K_{(i-1)} \leqslant i + e_{Ri},$$
$$i = 2, 3, \ldots, (n-1) \quad (9\text{-}23)$$

$$(i-1)K_1 + (i-2)K_2 + \cdots + K_{(i-1)} \geqslant i - d_{Ri},$$
$$i = 2, 3, \ldots, (n-1) \quad (9\text{-}24)$$

$$(i-1)K_1 + (i-2)K_2 + \cdots + K_{(i-1)} = i,$$
$$i = n, n+1, \ldots \quad (9\text{-}25)$$

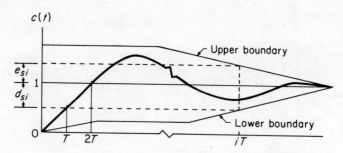

FIG. 9-4. Boundaries on the step response of the system [11].

Equation (9-25) ensures a zero steady-state sampled error and can be satisfied by requiring that both $1 - K(z)$ and its derivative contain the factor $(1 - z^{-1})$. Hence at $z = 1$, $K(z) = 1$ and $K'(z) = 0$. Using Eq. (9-9), the constraint equations are

$$\sum_{i=1}^{n} K_i = 1, \qquad \sum_{i=1}^{n} i K_i = 0 \qquad (9\text{-}26)$$

$$K_i = 0, \qquad i > n \qquad (9\text{-}27)$$

Equation (9-27) shows that the A matrix need be no larger than $[(n + 1) \times (n + 1)]$, while Eqs. (9-23), (9-24), and (9-26) now form a set of $2n - 2$ constraints.

Note that in cases where the plant is a nonminimum phase and/or unstable, it is desirable that $K(z)$ contain all the zeros of $G(z)$ outside the unit circle and/or $1 - K(z)$ contain all the poles outside the unit circle. If $G(z)$ is of the form

$$G(z) \frac{z^{-m}(1 - az^{-1})(h_0 h_1 z^{-1} + \cdots + h_s z^{-s})}{(1 - bz^{-1})(1 + g_1 z^{-1} + \cdots + g_r z^{-r})}, \qquad r \geqslant s \qquad (9\text{-}28)$$

where a and b represent a zero and a pole location outside the unit circle, then the appropriate constraint equations are obtained by setting $K(z) = 0$ and $z = a$ and by setting $1 - K(z) = 0$ at $z = b$:

$$\sum_{i=1}^{n} a^{n-i} K_i = 0, \qquad \sum_{i=1}^{n} b^{n-i} K_i = b^n \qquad (9\text{-}29)$$

It should also be noted that limiting n, the number of coefficients in $K(z)$, limits the size of matrix A and tends to limit the number of terms in the compensator $D(z)$.

The application of quadratic programming to statistical design of sampled-data control systems is now illustrated with two examples.

EXAMPLE 1

Referring to Fig. 9-2, the problem is to design a digital compensator $D(z)$ which will minimize the mean-square sampled error $e^2(nT)$ if

$$G(s) = \frac{1 + e^{-Ts}}{s^2(s+1)}, \quad T = .2$$

$$\phi_{nn}(\omega) = 0.1, \quad \phi_{ss}(\omega) = \frac{1}{1+\omega^2}, \quad \phi_{sn}(\omega) = \phi_{ns}(\omega) = 0$$

The solution obtained by an analytical approach has been presented by Tou [13]. His results were

$$K(z) = \frac{.556}{(z - .261)}$$
$$= .556z^{-1} + .145z^{-2} + .0379z^{-3} + .0099z^{-4} + \cdots$$

$$D(z) = \frac{29.25(z^2 - 1.819z + .819)}{(z^2 - .08z - .73)}$$

In general, preparing a problem for solution by quadratic programming requires the forming of the objective function and the writing of the linear constraint equations. In this example there are no linear constraint equations, and the objective function is given by Eq. (9-15). That is, the objective is

$$\text{Minimize } J = \hat{\mathbf{K}} \hat{\mathbf{A}} \hat{\mathbf{K}}^T$$

where $\hat{\mathbf{K}}$ is to be found by the computer program and $\hat{A}$ is given by Eq. (9-16). The elements of $\hat{A}$ are found by expanding $\phi_{ss}(z)$ and $\phi_{nn}(z)$ in a power series in z^{-1}. From a table of transforms,

$$\phi_{ss}(z) = \frac{-.2013z}{(z - .8019)(z - 1.221)}$$
$$= \frac{1}{2}\left(\sum_{i=1}^{-\infty} e^{iT}z^{-i} + 1 + \sum_{i=1}^{\infty} e^{-iT}z^{-i}\right)$$

$$\phi_{nn}(z) = .1$$

Using Eq. (9-13) to evaluate the elements of the 61×61 matrix $\hat{A}$ gives

$$\hat{A} = \begin{bmatrix}
 & 0 & 1 & 2 & 30 & 31 \cdots 60 \\
0 & .500000 & -.409366 & -.335160 & \cdots & .409366 \\
1 & -.409366 & .600000 & .409366 & \cdots & \\
2 & -.335160 & .409366 & .600000 & \cdots & \\
3 & -.274406 & .335160 & .409366 & \cdots & -A_{00} \\
 & \cdot & .274406 & .335160 & \cdots & \\
 & \cdot & \cdot & .274406 & \cdots & \\
 & \cdot & \cdot & \cdot & & \\
30 & -.001239 & \cdot & \cdot & .600000 & \\
31 & .409366 & & & & \\
 & \cdot & & & -A_{00} & A_{00} \\
60 & .001239 & & & &
\end{bmatrix}$$

This example was solved on an IBM 7040 General-Purpose Computer. The solution time was 2.28 min. The solution is tabulated as follows:

$$K_1 = .5548 \quad K_4 = .0102 \quad K_7 = .0002$$
$$K_2 = .1464 \quad K_5 = .0027 \quad K_8 = .0001$$
$$K_3 = .0386 \quad K_6 = .0007 \quad K_9 = K_{10} = \cdots$$
$$= K_{30} = 0$$
$$\overline{e^2 n(T)} = .2103$$

These values agree quite well with the series expansion of the solution obtained using the analytic approach. Note that a large majority of the elements in the vector $\hat{\mathbf{K}}$ are zero and that the corresponding $K(z)$ converges rapidly.

EXAMPLE 2

For the problem in Example 1 it is now required that the step response be bounded by the values given in Table 9-1 and that the steady-state error be zero by the eighth sampling instant. In Example 1, $\hat{A}$ was chosen to be a 61×61 matrix. In this example $\hat{A}$ is due to the constraint that the steady-state error to a step input goes to zero by the eighth sampling instant. There are 14 constraints of the form of Eq. (9-22) owing to the boundary values imposed on the step response at the first seven sampling instants and an additional constraint of the form of Eq. (9-21) owing to the requirement of zero steady-state error to the step input. Figure 9-5 illustrates the computer results. Note that the mean-square sampled error $\overline{e^2(nT)}$ has increased from .2103 in Example 1 to .2331 owing to the constraints.

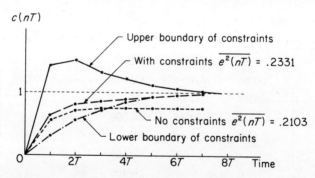

FIG. 9-5. Response of systems designed in Examples 1 and 2 to a unit step input [11].

In addition to the above constraints, Eqs. (9-26) and (9-27) are added to ensure a zero steady-state error in the ramp response. The computer solutions for $n = 6$ and $n = 8$ are illustrated in Figs. 9-6 and 9-7.

This work was further extended by Marowitz [14], who applied MP techniques in constrained problems of optimal filters design.

Table 9-1

BOUNDARY VALUES FOR THE STEP RESPONSE IN EXAMPLE 2

Sampling instant	ϵ_{si}	δ_{si}
T	.40	.70
$2T$	.50	.45
$3T$	.30	.30
$4T$	.20	.16
$5T$	.10	.08
$6T$	.05	.05
$7T$	.01	.01

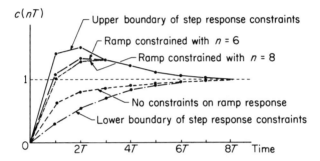

FIG. 9-6. Response of systems designed in Example 2 with and without a constraint on the ramp response [11].

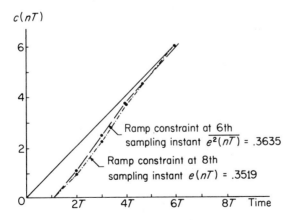

FIG. 9-7. Response of the systems designed in Example 2 to a unit ramp input [11].

9.4 APPLICATION IN OPTIMAL-ESTIMATION PROBLEMS [15]

Optimal estimation is basically an optimization problem, usually formulated as a minimization of mean-square error. Naturally, just as any other optimization problem, it can be formulated as a MP problem. A formulation was proposed by Issacs and Lochrie [15] for a recursive optimal-estimation problem subject to state inequality constraints. A NLP problem is formulated at each updating step of the recursive algorithm.

The system considered is described by the following dynamic state equations:

$$\dot{\mathbf{y}}(t) = A(t)\mathbf{y}(t) \tag{9-30}$$

where $\mathbf{y}(t)$ is an n-dimensional state vector and A is an $n \times n$ time-varying matrix.

The state transition equation between two subsequent times t_j, t_{j+1} ($t_j < t_{j+1}$) is given by

$$\mathbf{y}(j+1) = \Phi(j+1,j)\mathbf{y}(j) \tag{9-31}$$

where $\mathbf{y}(j) = \mathbf{y}(t_j)$ and $\Phi(j+1,j)$ is the state transition matrix, obtained by integrating the following differential equation:

$$\dot{\Phi}(t,j) = A(t)\Phi(t,j) \tag{9-32}$$

from $t = t_j$ to $t = t_{j+1}$ with the initial condition $\Phi(j,j) = I$.

The state variables are subject to a set of inequality constraints at all sampling times:

$$\mathbf{g}[\mathbf{y}(j)] \leqslant \mathbf{b} \tag{9-33}$$

where $\mathbf{g}$ is a convex vector function, $\mathbf{b}$ is a constant vector, and the set of inequalities in Eq. (9-33) defines a closed convex set in the n-dimensional space.

The measurement equations are given by

$$\mathbf{z}(j) = H(j)\mathbf{y}(j) + \mathbf{v}(j) \tag{9-34}$$

where $\mathbf{z}(j) = m$-dimensional measurement vector at $t = t_j$
$H(j) = m \times n$ measurement matrix at $t = t_j$
$\mathbf{v}(j) = m$-dimensional measurement noise vector, assumed to be zero-mean white noise

At each step, or sampling time $t = t_j$, an estimate of the state vector, $\hat{\mathbf{y}}(j)$ is obtained. The problem is to update the state estimate $\hat{\mathbf{y}}(j)$ using the measurement $\mathbf{z}(j)$ such that the squared observation errors are minimized subject to the constraints in Eq. (9-33).

The state $\mathbf{y}(j+1)$ is updated in a recursive fashion; i.e., given an estimate $\hat{y}(j)$ based on $z(1), \ldots, z(j)$, a new estimate $\hat{y}(j+1)$ is computed based only on $z(j+1)$ and $\hat{y}(j)$. To perform this, $\hat{y}(j+1)$ must be determined based on all the data, $z(1), \ldots, z(j+1)$. The expected measurement at $t = t_i$ is given by

$$\hat{z}(i) = H(i)\Phi(i, j+1)\hat{y}(j+1) \tag{9-35}$$

The following notation is introduced:

$$M_j = \begin{bmatrix} H(1)\Phi(1, j) \\ \vdots \\ H(j)\Phi(j, j) \end{bmatrix}, \quad \hat{z}_j^* = \begin{bmatrix} \hat{z}(1) \\ \vdots \\ \hat{z}(j) \end{bmatrix}, \quad z_j^* = \begin{bmatrix} z(1) \\ \vdots \\ z(j) \end{bmatrix} \tag{9-36}$$

where the $z(i)$'s ($i = 1, \ldots, j$) denote the vectors of actual measurements.
Considering all the measurements up to $t = t_{j+1}$, we have

$$\begin{bmatrix} M_j \Phi^{-1}(j+1, j) \\ H(j+1) \end{bmatrix} \hat{y}(j+1) = \begin{bmatrix} \hat{z}_j^* \\ \hat{z}(j+1) \end{bmatrix} \tag{9-37}$$

Equation (9-37) is not satisfied precisely since the measurements are corrupted by noise, v. Therefore, $y(j+1)$ is to be determined such that the following functional is minimized:

$$J_{j+1} = \left\| \begin{bmatrix} M_j \Phi^{-1}(j+1, j) \\ H(j+1) \end{bmatrix} \hat{y}(j+1) - \begin{bmatrix} \hat{z}_j^* \\ \hat{z}(j+1) \end{bmatrix} \right\|_{Q_{j+1}^*}^2 \tag{9-38}$$

subject to the constraints

$$g[\hat{y}(j+1)] \leqslant b \tag{9-39}$$

for all t considered.

The notation $\|x\|_{Q_{j+1}^*}$ denotes the quadratic form $x^T Q_{j+1}^* x$, where Q_{j+1}^* is a positive definite matrix. Assume that Q_{j+1}^* is of the form

$$Q_{j+1}^* = \begin{bmatrix} Q_j^* & 0 \\ 0 & Q_{j+1} \end{bmatrix} \tag{9-40}$$

Expanding the quadratic form in Eq. (9-38), we get

$$J_{j+1} = \|M_j \Phi^{-1}(j+1, j)\hat{y}(j+1) - z_j^*\|_{Q_j^*}^2 + \|H(j+1)\hat{y}(j+1) - z(j+1)\|_{Q_{j+1}}^2 \tag{9-41}$$

Expanding J_{j+1} eliminating terms not containing $\hat{y}(j+1)$ and denoting the remaining terms by the functional L_{j+1}, one obtains

$$L_{j+1} = \hat{y}^T(j+1)\{[\Phi^{-1}(j+1, j)]^T M_j^T Q_j^* M_j \Phi^{-1}(j+1, j) + H^T(j+1)Q_{j+1}H(j+1)\}\hat{y}(j+1) \\ - 2[z_j^{*T} Q_j^* M_j \Phi^{-1}(j+1, j) + z^T(j+1)Q_{j+1}H(j+1)]\hat{y}(j+1) \tag{9-42}$$

Now introduce the following abbreviating notations:

$$D_j = M_j^T Q_j^* M_j \qquad (9\text{-}43)$$

$$D_{j+1} = [\Phi^{-1}(j+1,j)]^T D_j \Phi^{-1}(j+1,j) + H^T(j+1)Q_{j+1}H(j+1) \qquad (9\text{-}44)$$

$$c_j^T = 2z_j^{*T} Q_j^* M_j \qquad (9\text{-}45)$$

$$c_{j+1}^T = c_j^T \Phi^{-1}(j+1,j) + 2z^T(j+1)Q_{j+1}H(j+1) \qquad (9\text{-}46)$$

Using Eqs. (9-42) through (9-46), one may formulate the following NLP problem:

$$\min\{\hat{y}^T(j+1)D_{j+1}\hat{y}(j+1) - c_{j+1}^T \hat{y}(j+1) \,|\, g[\hat{y}(j+1)] \leqslant b\} \qquad (9\text{-}47)$$

Consider now the computation of the matrix $\Phi^{-1}(j+1,j)$ appearing in Eqs. (9-44) and (9-46). For convenience, introduce the adjoint matrix defined by

$$\Psi^T(t, t_j) = \Phi^{-1}(t, t_j) \qquad (9\text{-}48)$$

The differential equation satisfied by the adjoint system can be obtained by noting that

$$\Psi^T(t, t_j)\Phi(t, t_j) = I \qquad (9\text{-}49)$$

and using Eq. (9-32),

$$\dot{\Psi}(t, t_j) = -A^T(t)\Psi(t, t_j) \quad \text{with } \Psi(t_j, t_j) = I \qquad (9\text{-}50)$$

The algorithm for solving the recursive estimation problem with constrained state variables is formulated as follows:

1. Compute $\Psi(j+1, j)$ by integrating Eq. (9-50) from $t = t_j$ to $t = t_{j+1}$.
2. Update the performance index L_{j+1}:
   (a)
   $$D_{j+1} = \Psi(j+1,j)D_j\Psi^T(j+1,j) + H^T(j+1)Q_{j+1}H(j+1)$$
   $$\text{with } D_0 = \{\text{cov}[\hat{y}(0)]\}^{-1} \qquad (9\text{-}51)$$
   where $\text{cov}(\mathbf{x}) = E[\mathbf{xx}^T]$, E denoting the mean or expected value.

   (b)
   $$c_{j+1} = \Psi(j+1,j)c_j + 2H^T(j+1)Q_{j+1}z(j+1)$$
   $$\text{with } c_0 = 2D_0\hat{y}(0) \qquad (9\text{-}52)$$

   (c)
   $$L_{j+1} = \hat{y}^T(j+1)D_{j+1}\hat{y}(j+1) - c_{j+1}^T\hat{y}(j+1) \qquad (9\text{-}53)$$

3. Solve the NLP problem
   $$\min\{L_{j+1} \,|\, g[\hat{y}(j+1)] \leqslant b\} \qquad (9\text{-}54)$$
   where the initial estimate is assumed to be $\Phi(j+1,j)\hat{y}(j)$.

9.5 IDENTIFICATION [16]

One of the most important problems in the design of control systems is that of determining the dynamic characteristics of the process or plant to be controlled. In this section, a method for identifying a linear system by measuring the input and output at a discrete number of points in time and then employing an iterative method which incorporates linear programming is presented [16].

The identification (parameter-estimation) problem consists of determining a mathematical model expressing the relationship between the sampled input and output sequences of a system. The model is formulated in terms of a rational z transform and is assumed to be linear and time invariant and have a single input and output.

The model, $G(z)$, is defined as a ratio of rational polynomials:

$$G(z) = \frac{N(z)}{D(z)} \tag{9-55}$$

where

$$N(z) = \alpha_0 + \alpha_1 z^{-1} + \cdots + \alpha_p z^{-p} \tag{9-56}$$

and

$$D(z) = 1 + \beta_1 z^{-1} + \cdots + \beta_q z^{-q} \tag{9-57}$$

The input and output sequences are time functions designated by $x(t)$ and $y(t)$, respectively, and are sampled at a uniform sampling rate T. These input and output sequences can be represented by z transforms:

$$\begin{aligned} X(z) &= x_0 + x_1 z^{-1} + \cdots + x_{n-1} z^{-(n-1)} \\ &= \sum_{j=0}^{n-1} x_j z^{-j} \end{aligned} \tag{9-58}$$

$$\begin{aligned} Y(z) &= y_0 + y_1 z^{-1} + \cdots + y_{n-1} z^{-(n-1)} \\ &= \sum_{j=0}^{n-1} y_j z^{-j} \end{aligned} \tag{9-59}$$

where $[x_n]$ and $[y_n]$ are the values of the input and output sequences at the sampling instants and n is the number of samples.

Figure 9-8 shows the plant and the model. The error between the model and the desired system (plant) response is

$$E(z) = \frac{X(z)N(z)}{D(z)} - Y(z) \tag{9-60}$$

$N(z)$ and $D(z)$ are to be selected such that

$$J = \sum_{j=0}^{n-1} e(jT) \tag{9-61}$$

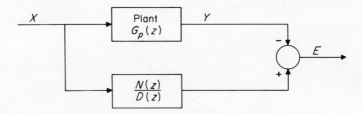

FIG. 9-8. The identification model [16].

is minimized. This will be accomplished by an iterative technique which incorporates linear programming.

Equation (9-60) can be approximated by [17]

$$E_i(z) = \frac{X(z)N_i(z)}{D_{i-1}(z)} - \frac{Y(z)D_i(z)}{D_{i-1}(z)} \qquad (9\text{-}62)$$

where i is the iteration number.

The division by the term $D_{i-1}(z)$ acts as a digital filter on the input and output (see Fig. 9-9) sequences, $[x_n]$ and $[y_n]$, yielding the filtered input and

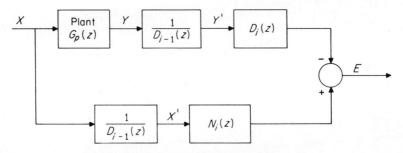

FIG. 9-9. Iterative identification model [16].

output sequences, $[x'_n]$ and $[y'_n]$, respectively. Thus Eq. (9-62) can be rewritten as

$$E_i(z) = X'_i(z)N_i(z) - Y'_i(z)D_i(z) \qquad (9\text{-}63)$$

where

$$X'_i(z) = \frac{X(z)}{D_{i-1}(z)}; \quad Y'_i(z) = \frac{Y(z)}{D_{i-1}(z)} \qquad (9\text{-}64)$$

The solution of Eq. (9-62) is initiated by assuming an initial solution for $D_0(z)$ and then, using Eq. (9-64), converting the original input and output sequences, $[x_n]$ and $[y_n]$, into the filtered input and output sequences, $[x'_n]$ and $[y'_n]$. These filtered sequences are then used in Eq. (9-63), yielding an expression in which the coefficients of $N_1(z)$, $D_1(z)$, and $E_1(z)$ are undetermined. This expression is a set of linear constraint equations representing the error between the model and the desired response at each sampling instant. Linear programming may be used to compute the values of the coefficients of $N(z)$

and $D(z)$ as defined by Eqs. (9-56) and (9-57) such that J in Eq. (9-61) is minimized.

The first application of linear programming will yield $D_1(z)$. This is then used to compute the new input and output sequences $[x'_n]$ and $[y'_n]$. These are again used in Eq. (9-63) to construct new constraint equations. Linear programming is again applied to yield $D_2(z)$ [and, incidentally, $N_2(z)$ and $E_2(z)$].

Convergence is attained when

$$D_i(z) = D_{i-1}(z) \tag{9-65}$$

Since Eq. (9-63) can be rewritten as

$$E_i(z) = \left[\frac{X(z)N_i(z)}{D_i(z)} - Y(z)\right]\frac{D_i(z)}{D_{i-1}(z)} \tag{9-66}$$

it is apparent that at convergence, Eq. (9-66) reduces to Eq. (9-60). Thus this iterative identification method is a means of providing a solution to the nonlinear computational problem as represented by Eq. (9-60) by repeatedly solving an associated linear problem.

The system error, as defined by Eq. (9-63), can be expressed in matrix form as

$$X'\mathbf{n} - Y'\mathbf{d} - \mathbf{e} = \mathbf{0} \tag{9-67}$$

where the X' and Y' matrices are

$$X' = \begin{bmatrix} x'_0 & 0 & \cdots & 0 \\ x'_1 & x'_0 & \cdots & 0 \\ \cdot & \cdot & & \cdot \\ \cdot & \cdot & & \cdot \\ \cdot & \cdot & & \cdot \\ x'_{n-1} & x'_{n-2} & \cdots & x'_{n-1-p} \end{bmatrix}$$

$$Y' = \begin{bmatrix} y'_0 & 0 & \cdots & 0 \\ y'_1 & y'_0 & \cdots & 0 \\ \cdot & \cdot & & \cdot \\ \cdot & \cdot & & \cdot \\ \cdot & \cdot & & \cdot \\ y'_{n-1} & y'_{n-2} & \cdots & y'_{n-1-q} \end{bmatrix} \tag{9-68}$$

and the $\mathbf{n}$, $\mathbf{d}$, and $\mathbf{e}$ column vectors are

$$\mathbf{n} = \begin{bmatrix} \alpha_0 \\ \alpha_1 \\ \cdot \\ \cdot \\ \cdot \\ \alpha_p \end{bmatrix}, \quad \mathbf{d} = \begin{bmatrix} 1 \\ \beta_1 \\ \cdot \\ \cdot \\ \cdot \\ \beta_q \end{bmatrix}, \quad \mathbf{e} = \begin{bmatrix} e_0 \\ e_1 \\ \cdot \\ \cdot \\ \cdot \\ e_{n-1} \end{bmatrix} \tag{9-69}$$

where the subscript i of Eq. (9-63) is implied.

The matrices X' and Y' are known for each iteration and the vectors $\mathbf{n}$, $\mathbf{d}$, and $\mathbf{e}$ are unknowns for which a solution is desired.

Since the vector $\mathbf{d}$ has a constant component, the following relationship can be defined:

$$\mathbf{d} = \mathbf{d}' + \mathbf{r}$$

where

$$\mathbf{d}' = \begin{bmatrix} 0 \\ \beta_1 \\ \cdot \\ \cdot \\ \cdot \\ \beta_q \end{bmatrix} \quad \text{and} \quad \mathbf{r} = \begin{bmatrix} 1 \\ 0 \\ \cdot \\ \cdot \\ \cdot \\ 0 \end{bmatrix} \quad (9\text{-}70)$$

Substituting Eq. (9-70) into Eq. (9-67) yields

$$X'\mathbf{n} - Y'(\mathbf{d}' + \mathbf{r}) - \mathbf{e} = 0 \quad (9\text{-}71)$$

Since $\mathbf{r}$ is a unit vector, Eq. (9-71) can be rewritten as

$$X'\mathbf{n} - Y'\mathbf{d}' - \mathbf{y}' - \mathbf{e} = 0 \quad (9\text{-}72)$$

where

$$\mathbf{y}' = Y'\mathbf{r} = \begin{bmatrix} y'_0 \\ y'_1 \\ \cdot \\ \cdot \\ \cdot \\ y'_{n-1} \end{bmatrix} \quad (9\text{-}73)$$

is the known output vector.

Since $\mathbf{y}'$ is a known vector, it is convenient to rewrite Eq. (9-72) as

$$X'\mathbf{n} - Y'\mathbf{d}' - \mathbf{e} = \mathbf{y}' \quad (9\text{-}74)$$

It is also desired that the unknowns, vectors $\mathbf{n}$, $\mathbf{d}'$, and $\mathbf{e}$, be unrestricted in sign. The unknown vectors can be expressed as (see Sec. 7.3)

$$\mathbf{n} = \mathbf{n}^+ - \mathbf{n}^-, \quad \mathbf{d}' = \mathbf{d}'^+ - \mathbf{d}'^-, \quad \text{and} \quad \mathbf{e} = \mathbf{e}^+ - \mathbf{e}^- \quad (9\text{-}75)$$

Substituting Eq. (9-75) into Eq. (9-74) yields

$$X'\mathbf{n}^+ - X'\mathbf{n}^- - Y'\mathbf{d}'^+ - Y'\mathbf{d}'^- - \mathbf{e}^+ + \mathbf{e}^- = \mathbf{y}' \quad (9\text{-}76)$$

representing the system at each sample instant. Since Eq. (9-76) is linear, linear programming can be used to generate solutions for the unknown coefficients.

Additional equations may be required to meet a variety of system requirements. For instance, it may be required that the system errors be constrained in a certain manner at specified sampling instants. Constraints of this type could appear as

$$-\xi'_j \leqslant e_j \leqslant \xi''_j \quad (9\text{-}77)$$

Since in most cases the error constraints are symmetrical, the problem is then to limit the absolute value of the error such that

$$|e_j| \leqslant \xi_j \tag{9-78}$$

Since the errors are unrestricted in sign, Eq. (9-78) becomes

$$e_j^+ - e_j^- \leqslant \xi_j \quad \text{and} \quad e_j^+ - e_j^- \geqslant -\xi_j \tag{9-79}$$

The inequalities can be rewritten as equations by the use of slack variables (see Chapter 3) and can be expressed in matrix form as

$$\begin{aligned} \mathbf{e}^+ - \mathbf{e}^- + \mathbf{s}' &= \boldsymbol{\xi} \\ -\mathbf{e}^+ + \mathbf{e}^- + \mathbf{s}'' &= \boldsymbol{\xi} \end{aligned} \tag{9-80}$$

where the vectors $\mathbf{s}'$, $\mathbf{s}''$, and $\boldsymbol{\xi}$ are nonnegative.

The conditions expressed by Eqs. (9-76) and (9-80) define the set of constraint equations used in the iterative identification method. Equation (9-76) holds for each sampling instant, while Eq. (9-80) holds only for those sample instants where limits are imposed upon the errors.

The fact that Eq. (9-76) holds for each sample instant limits the size of problems that can be solved using linear programming. However, it is not necessary that all of these constraint equations be used in the minimization problem.

The filtering of the input and output sequences as defined by Eq. (9-64) can be rewritten in the general form

$$A'(z) = D'_{i-1}(z) A(z) \tag{9-81}$$

where

$$\begin{aligned} D'_{i-1}(z) &= \frac{1}{D_{i-1}(z)} \\ &= 1 + d'_1(z) + \cdots + d'_n(z) + \cdots \end{aligned} \tag{9-82}$$

The quantity $D'_{i-1}(z)$ is of infinite length (but will be truncated in actual use) except for the trivial case where $D_{i-1}(z) = 1$. The input sequence $A(z)$ is defined as

$$A(z) = a_0 + a_1 z^{-1} + \cdots + a_{n-1} z^{-(n-1)} \tag{9-83}$$

When the filtered sequence $A'(z)$ is calculated using Eq. (9-81), the result is

$$A'(z) = a_0 + (a_0 d'_1 + a_1) z^{-1} + \cdots + (a_0 d'_{n-1} + \cdots + a_{n-1}) z^{-(n-1)} \tag{9-84}$$

It is evident from Eq. (9-84) that succeeding terms (higher powers of z^{-1}) have components from all past input data. Thus one can select those sample instants at which the response is to be matched most closely without losing information about the intermediate sampling instants.

The case for which this is not true is when $D_{i-1}(z) = 1$. In this case it is necessary to consider the transfer function for which the coefficients are to

be determined. This relationship is
$$A'(z)B(z) \tag{9-85}$$
where
$$A'(z) = a_0 + a'_1 z^{-1} + \cdots + a'_{n-1} z^{-(n-1)} \tag{9-86}$$
$$B(z) = b_0 + b_1 z^{-1} + \cdots + b_m z^{-m} \tag{9-87}$$
and
$$m < n \tag{9-88}$$

The transfer function coefficients (b's) are to be determined. Using Eqs. (9-86) and (9-87), Eq. (9-81) becomes

$$a'_0 b_0 + (a'_0 b_1 + a'_1 b_0)z^{-1} + \cdots + (a'_{n-1-m} b_m + a'_{n-m} b_{m-1} + \cdots + a'_{n-1} b_0)z^{-(n-1)} \tag{9-89}$$

It is evident from Eq. (9-89) that if more than $m+1$ equations are skipped, information will be lost. Since the solution $D_{i-1}(z) = 1$ could be computed, this limitation must be observed when selecting the spacing between the sampling instants to be matched.

Linear programming imposes the restriction that the objective function be linear. Most desired performance criteria can be obtained by using the general criterion of minimizing the sum of the weighted absolute error having the form

$$\text{Minimize } J = \sum_{j=0}^{n-1} c_j |e_j| \tag{9-90}$$

In matrix form Eq. (9-90) becomes

$$\text{Minimize } J = \mathbf{c}^T |\mathbf{e}| \tag{9-91}$$

where $\mathbf{c}$ is the weighting (cost) vector. This form of the performance criterion in Eq. (9-91) presents some difficulty. However, since the vectors $\mathbf{e}^+$ and $\mathbf{e}^-$ are not linearly independent, because they both appear in the objective function as pairs, the jth element of Eq. (9-91) is

$$|e_j^+ - e_j^-| = e_j^+ - e_j^- \tag{9-92}$$

where either e_j^+ or e_j^- or both are zero (see Sec. 7.3).

Substituting Eq. (9-92) into Eq. (9-91) yields

$$\text{Minimize } J = \mathbf{c}^T(\mathbf{e}^+ - \mathbf{e}^-) \tag{9-93}$$

The particular performance criterion used in this discussion is the sum of the absolute error (SAE), which is obtained by letting $\mathbf{c}$ be a unit scalar. Thus the performance criterion becomes

$$\text{Minimize } J = \mathbf{e}^+ - \mathbf{e}^- \tag{9-94}$$

This performance criterion in Eq. (9-94) used as the LP objective function, however, does not represent the true SAE of the problem being solved, the reason for this being that Eq. (9-94) generates the system error as represented by Fig. 9-9 and not the true system error represented by Fig. 9-8. The true SAE can be computed using the given input and output sequences, $[x_n]$ and $[y_n]$, and the computed coefficients.

Using the computed SAE, the problem is said to have converged when

$$\frac{|\text{SAE}_i - \text{SAE}_{i-1}|}{\text{SAE}_i} \leqslant V \tag{9-95}$$

where SAE_i and SAE_{i-1} are the present and previously calculated values of the SAE and V is an arbitrary limit.

Additional details concerning the computer program implementing this method are described in a thesis by Harrison [18]. This work was carried on by Burns [19], who developed a special computer package which realizes the described identification scheme along with an MP solution. The program is called Mathematical Programming Identification System—MPIS. It is written in FORTRAN IV, and it is implemented on the IBM 360 System of the University of Pennsylvania.

REFERENCES

1. M. Aoki, *Optimization of Stochastic Systems*, Academic Press, New York, 1967.
2. A. P. Sage, *Optimum Systems Control*, Prentice-Hall, Englewood Cliffs, N.J., 1968, chaps. 8–11.
3. H. W. Sorenson, "Kalman Filtering Techniques," *Advan. Control Systems*, **3**, pp. 219–292, C. T. Leondes, editor, Academic Press, New York, 1966.
4. R. S. Bucy and P. D. Joseph, *Filtering for Stochastic Processes with Applications to Guidance*, Wiley-Interscience, New York, 1968.
5. N. E. Nahi, *Estimation Theory and Applications*, Wiley, New York, 1969.
6. A. E. Bryson and Y. C. Ho, *Applied Optimal Control*, Blaisdell, Waltham, Mass., 1969, chaps. 10–14.
7. J. R. Fisher, "Optimal Nonlinear Filtering," *Advan. Control Systems*, **5**, pp. 197–300, C. T. Leondes, editor, Academic Press, New York, 1967.
8. D. Tabak, "Application of Mathematical Programming in the Design of Optimal Control Systems," Ph.D. Thesis, University of Illinois, Urbana, 1967.
9. D. Tabak, "Optimal Control of Nonlinear Discrete Time Systems by Mathematical Programming," *J. Franklin Inst.*, **289**, pp. 111–119, 1970.
10. B. C. Kuo, "Analysis and Synthesis of Sampled-Data Control Systems," Prentice-Hall, Englewood Cliffs, N.J., 1963.
11. A. J. Calise and K. A. Fegley, "Quadratic Programming in the Statistical Design of Sampled-Data Control Systems," *IEEE Trans. Automatic Control*, **AC-13**, pp. 77–80, 1968.

12. L. A. Wainstein and V. D. Zubakov, *Extraction of Signals from Noise*, R. A. Silverman, transl., Prentice-Hall, Englewood Cliffs, N.J., 1962, pp. 118–120.
13. J. T. Tou, "Statistical Design of Digital Control Systems," *IRE Trans. Automatic Control*, **AC-5**, pp. 290–297, 1960.
14. J. Marowitz, "Optimum Stochastic and Deterministic Control-System Synthesis Using Quadratic Programming," Ph.D. Thesis, University of Pennsylvania Philadelphia, 1969.
15. D. Isaacs and W. D. Lochrie, "Nonlinear Programming and Recursive Optimal Estimation," Allerton Conference of Circuits and Systems, Allerton, Ill., Oct. 1968.
16. R. W. Harrison and K. A. Fegley, "Identification of Linear Systems using Mathematical Programming," 1968 WESCON, Session 14, Los Angeles, Cal.
17. K. Steiglitz and L. E. McBride, "A Technique for the Identification of Linear Systems," *IEEE Trans. Automatic Control*, **AC-9**, pp. 461–464, 1965.
18. R. W. Harrison, "Identification of Linear Systems Using Mathematical Programming," M.S. Thesis, University of Pennsylvania, Philadelphia, May 1968.
19. J. F. Burns, "A Mathematical Programming Identification System (MPIS)," Ph.D. Thesis, University of Pennsylvania, Philadelphia, June 1969.

10

DISTRIBUTED-PARAMETER SYSTEMS

10.1 INTRODUCTION

Optimal control of distributed-parameter systems is one of the most active areas of current investigation and interest [1–4]. Many practical systems such as nuclear reactors [5] and space vehicles with flexible appendages [6, 7] are modeled as distributed-parameter systems. Considerable amounts of work in the development of theoretical and practical approaches to the optimal control of distributed-parameter systems were performed during the past few years. The literature cited here [8–17] is only a small percentage of that actually available. Extensive lists of references may be found in references 1–4. Several projects involving the application of mathematical-programming techniques in this area have been reported [18–20].

Sakawa [18] applied quadratic and linear programming to a problem involving a one-dimensional heat-conduction system. The highlights of his results are described in Sec. 10.2. This work was further extended by Lorchirachoonkul and Pierre [19, 20] to multivariable sampled-data distributed-parameter systems. Their approach is discussed in Sec. 10.3.

10.2 A HEAT-CONDUCTION SYSTEM

The heat-conduction system under consideration is described by the following partial differential equation [18]:

$$\frac{\partial^2 \tau(x, t)}{\partial x^2} = \frac{\partial \tau(x, t)}{\partial t} \qquad (10\text{-}1)$$

where $\tau(x, t)$ = Temperature distribution in the metal
x = Space coordinate, $0 \leqslant x \leqslant 1$
t = Time, $0 \leqslant t \leqslant T$

The time is normalized so that the coefficient corresponding to the thermal diffusivity in Eq. (10-1) is unity.

The initial and boundary conditions for the system are

$$\tau(x, 0) = 0 \qquad (10\text{-}2)$$

$$\left. \frac{\partial \tau(x, t)}{\partial x} \right|_{x=0} = h[\tau(0, t) - g(t)] \qquad (10\text{-}3)$$

$$\left. \frac{\partial \tau(x, t)}{\partial x} \right|_{x=1} = 0 \qquad (10\text{-}4)$$

where h = Heat-transfer coefficient, assumed to be constant
$g(t)$ = Temperature of the gas medium

The temperature of the gas medium, $g(t)$, is controlled by the fuel flow $u(t)$, and it is governed by the following differential equation:

$$\Theta \frac{dg(t)}{dt} + g(t) = u(t) \qquad (10\text{-}5)$$

where Θ is the time constant of the furnace.

The fuel flow $u(t)$ is the control variable of the problem. It controls the temperature distribution in the metal through the temperature of the gas medium. There is a certain desired temperature distribution along the metal $\tau_d(c)$. The purpose of the control action, at a given time $t = T$, is to minimize the functional

$$J = \int_0^1 [\tau_d(x) - \tau(x, T)]^2 \, dx \qquad (10\text{-}6)$$

The functional J is a measure of the deviation of the temperature distribution at $t = T$, $\tau(x, T)$, from the desired distribution, $\tau_d(x)$.

Taking the Laplace transform of Eq. (10-1) to (10-5), we obtain the system transfer function [18]:

$$P(x, s) = \frac{\tau(x, s)}{U(s)} = \frac{\cosh(1 - x)\sqrt{s}}{(\Theta s + 1)[(\sqrt{s}/h) \sinh \sqrt{s} + \cosh \sqrt{s}]} \qquad (10\text{-}7)$$

where $\tau(x, s) = L[\tau(x, t)]$
$U(s) = L[u(t)]$

According to the convolution theorem,

$$\tau(x, t) = \int_0^t p(x, \sigma) u(t - \sigma) \, d\sigma$$

$$= \int_0^t p(x, t - \sigma) u(\sigma) \, d\sigma \qquad (10\text{-}8)$$

where $p(x, t) = L^{-1}[P(x, s)]$.

At the final time $t = T$, we have

$$\tau(x, T) = \int_0^T p(x, T - \sigma)u(\sigma)\, d\sigma \tag{10-9}$$

The control variable $u(t)$ is also constrained as follows:

$$0 \leqslant u(t) \leqslant 1, \qquad 0 \leqslant t \leqslant T \tag{10-10}$$

Before applying any MP technique, the integrals in Eq. (10-6) and (10-9) are discretized. Simpson's composite formula is used [21], and the integral in Eq. (10-6) becomes

$$J \simeq \sum_{i=0}^{N} c_i [\tau_d(x_i) - \tau(x_i, T)]^2 \tag{10-11}$$

where

$$x_i = \frac{i}{N}, \qquad i = 0, 1, \ldots, N$$

$$N = \text{even number}$$

$$c_0 = c_N = \frac{1}{3N}$$

$$c_1 = c_3 = \cdots = c_{N-1} = \frac{4}{3N} \tag{10-12}$$

$$c_2 = c_4 = \cdots = c_{N-2} = \frac{2}{3N}$$

Applying the same integration scheme, the integral in Eq. (10-9) becomes

$$\tau(x_i, T) \simeq T \sum_{j=0}^{N} c_j p(x_i, T - \sigma_j) u(\sigma_j) \tag{10-13}$$

where $\sigma_j = \frac{jT}{N}, j = 0, 1, \ldots, N$.

Introducing the notation

$$a_{ij} = t c_j p(x_i, T - \sigma_j)$$
$$u_j = u(\sigma_j) \tag{10-14}$$
$$\tau_{di} = \tau_d(x_i)$$

and substituting Eq. (10-14) into Eq. (10-13) and then into Eq. (10-11), we obtain

$$J \simeq \sum_{i=0}^{N} c_i \left(\tau_{di} - \sum_{j=0}^{N} a_{ij} u_j \right)^2 \tag{10-15}$$

The constraints in Eq. (10-10) are written as

$$0 \leqslant u_j \leqslant 1, j = 0, 1, \ldots, N \tag{10-16}$$

Now we have a classic MP problem with $N + 1$ variables u_j and $N + 1$ inequality constraints in Eq. (10-16). Since J is quadratic and the constraints are linear, we have a QP problem. The same problem can also be formulated as a LP problem using the following performance index:

$$J \simeq \sum_{i=0}^{N} c_i \left| \tau_{di} - \sum_{j=0}^{N} a_{ij} u_j \right| \tag{10-17}$$

The absolute value expression in Eq. (10-17) is then treated using the techniques described in Chapters 5 and 7.

10.3 MULTIVARIABLE DISTRIBUTED-PARAMETER SAMPLED-DATA SYSTEMS

Lorchirachoonkul and Pierre extended the work of Sakawa to multiple-input distributed-parameter systems with sampled data [19, 20]. The basic configuration of the system considered is shown in Fig. 10-1. The input to

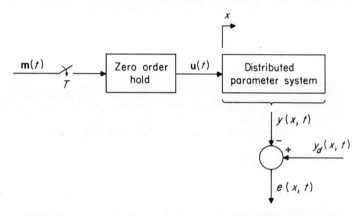

FIG. 10-1. Sampled-data representation for distributed-parameter systems.

the system is the r-dimensional control vector $\mathbf{m}(t)$. Each component of the control vector passes through a uniform sampler of period T and through a zero-order hold, thus producing the r-dimensional input vector $\mathbf{u}(t)$. The distributed-parameter system is single dimensional and is characterized by an output function $y(x, t)$. The output depends on the space coordinate x and on time t. A desired output $y_d(x, t)$ is specified. The error between the desired and the actual output is

$$e(x, t) = y_d(x, t) - y(x, t) \tag{10-18}$$

The purpose of the control action is to minimize this error. This consideration dictates the choice of the performance index. The space coordinate x is divided into N_x discrete intervals. The total time interval considered is NT.

The output is characterized by [20]

$$y(x, t) = k(x, t) + \int_0^t \mathbf{g}^T(x, t, \tau)\mathbf{u}(\tau)\, d\tau \tag{10-19}$$

where $k(x, t)$ = contribution of the initial distribution $y(x, 0)$ to the response of the system

$\mathbf{g}^T(x, t, \tau)$ = r-dimensional row vector relating the controlling action to the output distribution $y(x, t)$.

At the end of the jth sampling instant, the output can be expressed as

$$y(x, jT) = k(x, jT) + \sum_{i=1}^{j} \mathbf{h}_j^T(x, iT)\mathbf{m}(i-1) \tag{10-20}$$

where
$jT = \lim_{\varepsilon \to 0^+} (jT - \varepsilon)$
$\mathbf{m}(i-1) = \mathbf{m}[(i-1)T]$
$\mathbf{h}_j^T(x, iT)$ = r-dimensional row vector given by

$$\mathbf{h}_j^T(x, iT) = \int_0^T \mathbf{g}^T[x, jT, (i-1)T + \tau]\, d\tau \tag{10-21}$$

At the end of N sampling periods, the following set of equations is obtained:

$$\mathbf{Y}(x) = \mathbf{K}(x) + H(x)\mathbf{M} \tag{10-22}$$

where

$$\mathbf{Y}(x)_{N\times 1} = \begin{bmatrix} y(x, T) \\ y(x, 2T) \\ \cdot \\ \cdot \\ \cdot \\ y(x, NT) \end{bmatrix} \quad \mathbf{K}(x)_{N\times 1} = \begin{bmatrix} k(x, T) \\ k(x, 2T) \\ \cdot \\ \cdot \\ \cdot \\ k(x, NT) \end{bmatrix} \quad \mathbf{M}_{rN\times 1} = \begin{bmatrix} \mathbf{m}(0) \\ \mathbf{m}(1) \\ \cdot \\ \cdot \\ \cdot \\ \mathbf{m}(N-1) \end{bmatrix}$$

$$H(x)_{N\times Nr} = \begin{bmatrix} \mathbf{h}_1^T(x, T) & \mathbf{0}^T & \cdots & \mathbf{0}^T \\ \mathbf{h}_2^T(x, T) & \mathbf{h}_2^T(x, 2T) & \cdots & \mathbf{0}^T \\ \cdot & \cdot & & \cdot \\ \cdot & \cdot & & \cdot \\ \cdot & \cdot & & \cdot \\ \mathbf{h}_N^T(x, T) & \mathbf{h}_N^T(x, 2T) & \cdots & \mathbf{h}_N^T(x, NT) \end{bmatrix}$$

$\mathbf{0}^T = [0, 0, \ldots, 0]$ = r-dimensional zero row vector

The performance index is of the minimum absolute value of the error type:

$$\text{Minimize } J = \sum_{i=1}^{N_x} \sum_{j=1}^{N} a_{ij}\, |e(x_i, jT)| \tag{10-23}$$

The error function e is expressed as a function of the control vector $\mathbf{M}$ using Eqs. (10-18) and (10-22). The absolute value in the performance index is converted to a linear form using methods described in Chapters 5 and 7. In addition, the control variable is constrained as follows:

$$0 \leqslant m_j(i) \leqslant m_{ji\max}, \quad \begin{array}{l} i = 0, 1, \ldots, N-1 \\ j = 1, 2, \ldots, r \end{array} \tag{10-24}$$

Under certain circumstances, intersampling ripple may appear in the optimal distribution. By proper selection of the bounds on the derivatives of the error, ripple can be effectively restrained.

Differentiating Eq. (10-19) with respect to time, we have

$$\frac{\partial y(x, t)}{\partial t} = \frac{\partial k(x, t)}{\partial t} + \int_0^t \frac{\partial \mathbf{g}^T(x, t, \tau)}{\partial t} u(\tau)\, d\tau + \mathbf{g}^T(x, t, \tau) u(t) \tag{10-25}$$

The jth partial derivative with respect to t can be written

$$\frac{\partial^j y(x, t)}{\partial t^j} = \frac{\partial^j k(x, t)}{\partial t^j} + \int_0^t \frac{\partial^j \mathbf{g}^T(x, t, \tau)}{\partial t^j} u(\tau)\, d\tau$$
$$+ \sum_{i=1}^{j} \frac{\partial^{j-i}}{\partial t^{j-i}} \left[u(t) \frac{\partial^{i-1} \mathbf{g}^T(x, t, \tau)}{\partial t^{i-1}} \bigg|_{\tau=t} \right] \tag{10-26}$$

At $t = iT$ we obtain

$$\frac{\partial^j y(x, iT)}{\partial t^j} = \frac{\partial^j k(x, iT)}{\partial t^j} + \int_0^{iT} \frac{\partial^j \mathbf{g}^T(x, iT, \tau)}{\partial t^j} u(\tau)\, d\tau$$
$$+ u(t) \sum_{i=1}^{j} \frac{\partial^{j-i}}{\partial t^{j-i}} \left[\frac{\partial^{i-1} \mathbf{g}^T(x, t, \tau)}{\partial t^{i-1}} \bigg|_{\tau=t} \right]\bigg|_{t=iT} \tag{10-27}$$

Similarly, for the jth partial derivative with respect to x,

$$\frac{\partial^j y(x, t)}{\partial x^j} = \frac{\partial^j k(x, t)}{\partial x^j} + \int_0^t \frac{\partial^j \mathbf{g}^T(x, t, \tau)}{\partial x^j} u(\tau)\, d\tau \tag{10-28}$$

The jth partial derivatives with respect to t and x of the error response evaluated at $x = x_m$ and $t = iT$ are defined as follows:

$$e_t^j(x_m, iT) = \frac{\partial^j y_d(x_m, iT)}{\partial t^j} - \frac{\partial^j y(x_m, iT)}{\partial t^j} \tag{10-29}$$

$$e_x^j(x_m, iT) = \frac{\partial^j y_d(x_m, iT)}{\partial x^j} - \frac{\partial^j y(x_m, iT)}{\partial x^j} \tag{10-30}$$

To restrain intersampling ripple, the absolute values of the time and spatial derivatives of the error response are bounded by some constants. These constants are determined by the designer on the basis of the degree of ripple that can be tolerated in the optimal system.

Thus we have the following inequality constraints:

$$|e_t^j(x_m, iT)| \leqslant E_{t, mi}^j \tag{10-31}$$

$$|e_x^j(x_m, iT)| \leqslant E_{x, mi}^j \tag{10-32}$$

The absolute value expressions in the constraints in Eqs. (10-31) and (10-32) are converted into linear expressions following the methods described in Chapters 5 and 7.

EXAMPLE

To illustrate the method described in this section, the system characterized by

$$\frac{\partial y(x,t)}{\partial t} = \frac{\partial^2 y(x,t)}{\partial x^2}, \quad t > 0, \quad 0 \leqslant x \leqslant 1 \tag{10-33}$$

with the boundary conditions

$$y(0, t) = y(1, t) = u(t) \tag{10-34}$$

and

$$y(x, 0) = 0 \tag{10-35}$$

is used. The boundary input $u(t)$ is to be determined such that the performance index, with all a_{ij}'s equal to unity in Eq. (10-23) is minimized. The desired distribution $y_d(x, t)$ equals 1, and the constraint on the controlling action is

$$0 \leqslant m_j \leqslant 2 \tag{10-36}$$

The solution to the partial differential equation (10-33) with the boundary conditions in Eqs. (10-34) and (10-35) can be written as

$$y(x, t) = 4 \int_0^t \sum_{n=1}^\infty w_{2n-1} \sin(w_{2n-1}x) \exp[-w_{2n-1}^2(t-\tau)] u(\tau)\, d\tau \tag{10-37}$$

where

$$w_{2n-1} = (2n-1)\pi \tag{10-38}$$

From the theory of Fourier's series, it is known that at $x = 0$ and $x = 1$, the infinite series in the right-hand member of Eq. (10-37) does not converge to $y(x, t)$. However, from the boundary conditions in Eq. (10-34), the values of the output distribution at the boundaries are known to equal the input $u(t)$.

Then, on the basis of Eq. (10-21), each nonzero element of $H(x)$ is

$$h_j(x, iT) = 4 \sum_{n=1}^\infty \frac{\sin(w_{2n-1}x)}{w_{2n-1}} \exp[-w_{2n-1}^2(j-i+1)T][\exp(w_{2n-1}^2 T) - 1] \tag{10-39}$$

From Eq. (10-20), the distributed output is

$$y(x, jT) = \sum_{i=1}^j m_{i-1} h_j(x, iT) \tag{10-40}$$

The general term in the error sequence is then, from Eq. (10-18),

$$e(x_m, jT) = 1 - \sum_{i=1}^j m_{i-1} h_j(x_m, iT) \tag{10-41}$$

From the boundary conditions given in Eqs. (10-34) and (10-35), the output distribution of the system is known to be symmetrical: Thus characteristic values of x need be specified either in the interval defined by $0 \leqslant x \leqslant .5$ or in the interval defined by $.5 \leqslant x \leqslant 1$, but not in both. In this example, N_x is set at 3 with $x_1 = .1$, $x_2 = .3$, and $x_3 = .5$. Finally, the sampling period T is assumed to be .01 sec, and the total number of samples of interest is 30.

The optimal input, computed by use of the IBM 1620-1311 Linear Programming System, is shown in Fig. 10-2 and listed in Table 10-1. The optimal output is computed by use of Eq. (10-40) and is plotted as shown in Fig. 10-3 and tabulated in Table 10-2. The desired distribution as seen from Fig. 10-3 can be obtained in less than .15 sec for all practical purposes.

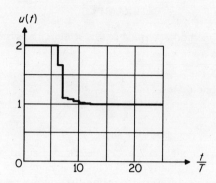

FIG. 10-2. Optimal input which results in the minimization of the performance index with all $a_{ij} = 1$, $m_{ji_{max}} = 2.0$, and $T = .01$ sec.

Table 10-1

OPTIMAL-CONTROL SEQUENCE WHICH RESULTS IN THE MINIMIZATION OF THE PERFORMANCE MEASURE, WITH ALL a_{ij}'s $= 1$, $m_{ji_{max}} = 2.0$, AND $T = .01$ sec

Time (sec)	Control sequence
.00	2.000
.01	2.000
.02	2.000
.03	2.000
.04	2.000
.05	2.000
.06	1.665
.07	1.091
.08	1.089
.09	1.045
.10	1.025
.11	1.013
.12	1.007
.13	1.004
.14	1.002
.15	1.001
.16	1.001
.17	.999
.18	1.000
.19	1.000
.20	1.000
.21	1.000
.22	1.000
.23	1.000
.24	1.000
.25	1.000
.26	1.000
.27	1.000
.28	1.000
.29	1.000

Sec. 10.3   Multivariable Distributed-Parameter Sampled-Data Systems   **227**

Table 10-2

OUTPUT RESPONSE OF THE SYSTEM ASSOCIATED WITH THE OPTIMAL INPUT IN TABLE 10-1

Distributed output, $y(x, t)$

j	$x = .1$	$x = .2$	$x = .3$	$x = .4$	$x = .5$
.01	.9590	.3146	.0678	.0094	.0016
.02	1.2342	.6347	.2682	.0964	.0497
.03	1.3666	.8306	.4499	.2335	.1649
.04	1.4501	.9683	.6043	.3824	.3084
.05	1.5115	1.0767	.7393	.5273	.4554
.06	1.5614	1.1682	.8592	.6628	.5956
.07	1.4436	1.1956	.9557	.7857	.7253
.08	1.1602	1.1231	1.0000	.8819	.8357
.09	1.0920	1.0624	1.0000	.9371	.9109
.10	1.0492	1.0339	1.0000	.9660	.9519
.11	1.0268	1.0183	1.0000	.9817	.9741
.12	1.0143	1.0099	1.0000	.9901	.9860
.13	1.0077	1.0077	1.0000	.9947	.9925
.14	1.0042	1.0029	1.0000	.9971	.9959
.15	1.0022	1.0016	1.0000	.9985	.9978
.16	1.0011	1.0008	1.0000	.9992	.9988
.17	1.0008	1.0005	1.0000	.9995	.9993
.18	.9998	1.0001	1.0000	.9998	.9997
.19	.9999	.9999	.9999	.9998	.9998
.20	.9999	.9999	.9999	.9998	.9998
.21	.9999	.9999	.9999	.9999	.9999
.22	.9999	.9999	.9999	.9999	.9999
.23	.9999	.9999	.9999	.9999	.9999
.24	.9999	.9999	.9999	.9999	.9999
.25	.9999	.9999	.9999	.9999	.9999
.26	.9999	.9999	.9999	.9999	.9999
.27	.9999	.9999	.9999	.9999	.9999
.28	.9999	.9999	.9999	.9999	.9999
.29	.9999	.9999	1.0000	.9999	.9999
.30	.9999	.9999	1.0000	.9999	.9999

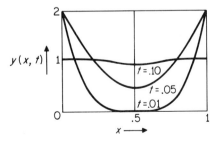

FIG. 10-3. Output response of the system associated with the optimal input in Table 10-2.

REFERENCES

1. M. Athans, "Some Remarks on the Control of Distributed Parameter Systems," 1969 JACC, Boulder, Colo.

2. S. K. Mitter, "Optimal Control of Distributed Parameter Systems," 1969 JACC, Boulder, Colo.

3. P. K. C. Wang, "Mathematical Modelling of Systems with Distributed Parameters," 1969 JACC, Boulder, Colo.

4. J. C. Willems, "A Survey of Stability of Distributed Parameter Systems," 1969 JACC, Boulder, Colo.

5. D. L. Briggs and C. N. Shen, "Distributed Parameter Optimum Control of a Nuclear Rocket with Thermal Stress Constraints," *Trans. ASME, J. Basic Eng.*, pp. 300–306, June 1967.

6. N. N. Puri and D. Tabak, "Analysis and Control of Flexible Structures," Princeton Conference of Information Sciences and Systems, Princeton, N.J., March 1968, pp. 237–241.

7. D. Tabak, "Analysis and Control of Space Vehicles with Flexible Appendages," Technical Report, Wolf R&D Corp., Riverdale, Md., Aug. 1968.

8. P. K. C. Wang, "Control of Distributed Parameter Systems," *Advances in Control Theory*, **1**, edited by C. T. Leondes, pp. 75–172, Academic Press, New York, 1964.

9. A. G. Butkovskii, "Optimal Control Theory of Distributed Parameter Systems," Nauka, Moscow, 1965. (English translation to appear.)

10. W. L. Brogan, "Optimal Control Theory Applied to Systems Described by Partial Differential Equations," *Advances in Control Theory*, **6**, edited by C. T. Leondes, pp. 221–316, Academic Press, New York, 1968.

11. M. H. Yeh and J. T. Tou, "Optimum Control of a Class of Distributed Parameter Systems," *IEEE Trans. Automatic Control*, **AC-12**, pp. 29–37, 1967.

12. H. Erzberger and M. Kim, "Optimum Boundary Control of Distributed Parameter Systems," *Inform. Control*, **9**, pp. 265–278, 1966.

13. A. I. Egorov, "On Optimal Control of Processes in Distributed Objects," *J. Appl. Math. Mech.*, **27**, pp. 688–698, 1963.

14. Y. Yavin and R. Sivan, "The Optimal Control of a Distributed Parameter System," *IEEE Trans. Automatic Control*, **AC-12**, pp. 758–761, 1967.

15. V. Komkov, "The Optimal Control of a Transverse Vibration of Beam," *SIAM J. Control*, **6**, pp. 401–421, 1968.

16. A. P. Sage, "Gradient and Quasilinearization Computational Techniques for Distributed Parameter Systems," *Intern. J. Control*, **6**, pp. 81–98, 1967.

17. E. I. Axelband, "An Approximation Technique for the Optimal Control of Linear Distributed Parameter Systems with Bounded Inputs," *IEEE Trans. Automatic Control*, **AC-11**, pp. 42–45, 1966.

18. Y. Sakawa, "Solution of an Optimal Control Problem in a Distributed-Parameter System," *IEEE Trans. Automatic Control*, **AC-9**, pp. 420–426, 1964.

19. V. Lorchirachoonkul, "Optimal Sampled-Data Control of Distributed-Parameter Systems," Ph.D. Thesis, Montana State University, Bozeman, March 1967.

20. V. Lorchirachoonkul and D. A. Pierre, "Optimal Control of Multivariable Distributed-Parameter Systems through Linear Programming," 1967 JACC, pp. 702–710, Philadelphia, Pa.

21. A. D. Booth, "Numerical Methods," Butterworth, London, 1955.

AUTHOR INDEX

A

Abadie, J., 23, 57, 68
Aoki, M., 217
Ash, M., 110, 113, 133
Athans, M., 7, 68, 228
Axelband, E. I., 229

B

Balinski, M. L., 24
Barr, R. D., 110
Beale, E. M. L., 57
Beightler, C. S., 7, 24, 57, 64, 69, 134
Bellman, R., 4, 8, 133
Benders, J. F., 24
Blum, S., 170
Boltyanskii, V. G., 3, 7, 69, 195
Boot, J. C. G., 24, 57
Booth, A. D., 229
Brentani, P. B., 64, 69
Briggs, D. L., 228
Brockstein, A. J., 172, 195
Brogan, W. L., 228
Bryson, A. E., 8, 217
Bucy, R. S., 217
Burns, J. F., 217, 218
Butkovskii, A. G., 228

C

Calise, A. J., 217
Canon, M. D., 170, 175
Carroll, C. W., 50, 57
Chang, S. S. L., 195
Clohessy, W. H., 80, 110
Collatz, L., 133
Courant, R., 68, 117, 134
Crandall, S. H., 171
Cullum, C. D., 175, 196
Cullum, J., 68, 69
Cutler, L., 138, 170

D

Dantzig, G. B., 24, 27, 56, 94, 110
Davidon, W. C., 54, 58, 118, 134
Denham, W. F., 8
Desoer, C. A., 110
Djadjuri, K., 158, 170
Duffin, R. J., 24

E

Eaton, J. H., 170
Egorov, A. I., 228
Enns, M., 170

Erzberger, H., 228
Eveleigh, V. W., 8

F

Falb, P. L., 7, 68
Fan, L. T., 195
Fath, A. F., 78, 86, 109, 110
Fegley, K. A., 141, 170, 196, 217, 218
Fiacco, A. V., 24, 50, 53, 57, 131, 178, 185, 196
Fisher, J. R., 217
Fletcher, R., 54, 58, 118, 134
Fomin, S. V., 8
Frank, M., 40, 57
Fromovitz, S., 175, 196

G

Gamkrelidze, R. V., 3, 7, 69, 195
Gass, S. I., 24, 57
Gelfand, I. M., 8
Gomory, R. E., 171
Graves, R. L., 24

H

Hadley, G., 23, 24, 56, 57
Halkin, H., 63, 69, 195
Hanson, M. A., 63, 69
Harrison, R. W., 217, 218
Hestenes, M. R., 58
Higgins, T. J., 78, 109
Ho, Y. C., 64, 69, 217
Holtzman, J. M., 195
Hsu, M. I., 170
Hurwicz, L., 63, 68

I

Isaacs, D., 8, 208, 218

J

Jizmagian, G. S., 94, 97, 110
Jordan, B. W., 195
Joseph, P. D., 217
Jury, E. I., 182, 196

K

Katz, S., 195
Kelley, H. J., 58
Kenneth, P., 8
Kim, M., 158, 170, 228
Komkov, V., 228
Kopp, R. E., 8
Kowalik, J., 134
Krelle, W., 23, 57
Kuhn, H. W., 19, 24, 38, 49, 63, 174, 196
Kunzi, H. P., 23, 57
Kuo, B. C., 69, 110, 169, 170, 172, 195, 196, 217

L

Lack, G. N. T., 170
Larson, R. E., 6, 8
Lasdon, L. S., 58, 131, 134
Lee, B. W., 182, 196
Lee, E. B., 7, 68
Leitmann, G., 8
Llewellyn, R. W., 24, 57
Lochrie, W. D., 208, 218
Lorchirachoonkul, V., 219, 222, 229
Luenberger, D. G., 8, 132, 134

M

Mangasarian, O. L., 175, 196
Markus, L., 7, 68
Marowitz, J., 206, 218
McBride, L. E., 218
McCormick, G. P., 24, 50, 53, 57, 131, 178, 185, 196
McGill, R., 8
Meyer, R., 132, 134, 195, 196
Mishchenko, E. F., 7, 69, 195
Mitter, S. K., 58, 134, 228
Mohler, R. R., 110
Moyer, H. G., 8
Myers, G. E., 57, 58

N

Nahi, N. E., 217
Narendra, K. S., 58

Neustadt, L. W., 63, 69
Norris, D. O., 69

O

Oettli, W., 23, 57
Osborne, M. R., 134

P

Pearson, J. B., 175, 196
Pearson, J. D., 52, 54, 57, 195
Peterson, E. L., 24
Pierre, D. A., 219, 222, 229
Pierson, B. L., 117, 133
Plant, J. B., 8
Polak, E., 169, 171, 175, 195, 196
Ponstein, J., 69
Pontryagin, L. S., 3, 7, 63, 69, 195
Popov, V. M., 121, 134
Porcelli, G., 141, 170, 196
Powell, M. J. D., 54, 58
Price, H. J., 110
Propoi, A. I., 141, 170, 195
Pshenichniy, B. N., 69
Puri, N. N., 228

R

Reeves, C. M., 118, 134
Rice, R. K., 131, 134
Roberts, J. J., 114, 133
Rosen, J. B., 46, 49, 57, 110, 132, 134, 172, 175, 193, 195, 196
Rosztoczy, Z. R., 133
Rozonoer, L. I., 195
Russell, D. L., 69

S

Sage, A. P., 8, 68, 217, 228
Sakawa, Y., 219, 222, 229
Scarborough, J. B., 69
Scharmack, D. K., 8
Shen, C. N., 228
Simonnard, M., 24, 56
Sinnott, J. F., 132, 134

Sivan, R., 228
Smith, H. P., 114, 133
Sorenson, H. W., 217
Sridhar, R., 175, 196
Steiglitz, K., 218

T

Tabak, D., 58, 69, 110, 133, 134, 170, 196, 217, 228
Torng, H. D., 137, 139, 170
Tou, J. T., 172, 195, 205, 218, 228
Tripathi, S. S., 58
Tucker, A. W., 19, 24, 38, 49, 63, 174, 196

V

Vajda, S., 23
Valentine, F. A., 24, 57
Van Slyke, R. M., 69, 94, 110
Volz, R. A., 75, 110

W

Waespy, C. M., 79, 94, 109
Wainstein, L. A., 218
Wang, C. S., 195
Wang, P. K. C., 228
Waren, A. D., 58, 131, 134
Weaver, L. E., 133
Whalen, B. H., 6, 8, 136, 149, 170
Wilde, D. J., 7, 24, 57, 64, 69, 134
Willems J. C., 228
Wiltshire, R. S., 80, 110
Wolfe, P., 24, 39, 40, 57, 157

Y

Yavin, Y., 228
Yeh, M. H., 228

Z

Zadeh, L. A., 6, 8, 110, 170
Zener, C. M., 24
Zoutendijk, G., 24, 43, 45, 46, 57
Zubakov, V. D., 218

SUBJECT INDEX

A

Adams interpolation, 117
Adjoint variables, 3
Admissible, control functions, 86
Admissible, trajectories, 86
Approximation, first-order, 117
Approximation, higher, 117
Artificial variables, 139
Attainability, set of, 86

B

Basic feasible solution, 27, 30
Basic solution, 27, 40
Basic variables, 27, 31
Bolza problem, 3

C

Calculus of variations, 2, 3
Computer control, 121, 182
Concave function, 16
Conjugate direction methods, 54
Conjugate gradient methods, 118, 131
Constraint qualification, 22
Constraints, 1, 9, 26, 61
 equality, 10, 26
 inequality, 10, 26, 114, 122

Control optimal, 1
Control systems, 1
Control variables, 2, 113
Convex function, 16
Convex set, 15, 16, 27, 208
Convexity, 15, 38
Costate variables, 3
Costate vector, 4
Cramer's rule, 163
Crout's algorithm, 163

D

Digital controller, design, 141, 181, 184, 187, 197, 200
Digital filter, 212
Direct digital control, 131, 182
Direction matrix, 43
Discrete-data systems, 4, 25
Discretization, 67, 117, 132
Distributed-parameter systems, 219
Dual problem, 21, 91
Duality, 20, 91
Dynamic programming, 2, 4, 113, 163, 172
Dynamic programming, state increment algorithm, 6
Dynamic programming, storage requirements, 5

E

Estimation, optimal, 197, 208
Euler-Language equation, 3
Extreme point, 28

F

Feasible directions, method, 43
Feasible region, 11, 13, 41
Feasible solution, 9, 27
Fibonacci search, 118
Fletcher-Powell-Davidon algorithm, 54
Flux, neutron, 112
Frank and Wolfe method in QP, 40
Fuel optimal problem, 77

G

Generalized programming, 72, 94
Geometric programming, 14
Gradient, 42
Gradient, first-order methods, 53
Gradient, large-step techniques, 43
Gradient, second-order methods, 53
Gradient, small step techniques, 43
Gradient approach to NLP, 41
Gradient projection, 49
Gradient projection method, 46

H

Hamilton-Jacobi-Bellman theory, 4
Hamiltonian, 3, 4, 174
Heat-conduction system, 219
Heaviside unit function, 117
Hessian matrix, 53
Hyperplane, 27, 46
Hyperplane, supporting, 45

I

Identification, 197, 211
Ill-conditioning effects, 169
Infinite solution, LP problem, 32
Integer programming, 15, 160

K

Kuhn-Tucker Conditions, 20, 49, 64
Kuhn-Tucker Theorem, 19, 38, 63, 174

L

Lagrange function, 19, 174
Lagrange multipliers, 3, 19, 174
Lagrange problem, 2, 3, 4
Linear programming, 12, 25, 70, 82, 91, 132, 136, 141, 169, 211, 222
Linear systems, continuous-time, 70
Linear systems, discrete-time, 135, 169

M

Manifold, linear, 46
Master program, restricted, 95
Mathematical programming, 6, 9, 59, 62, 70, 111, 172
Mayer problem, 2, 3
Maximum principle, 2, 3, 64, 132, 172
Metric, 43, 52
Minimax problem, 113
Minimum, global, 18
Minimum, local, 18
Minimum energy problem, 71
Minimum error problem, 140, 160
Minimum fuel problem, 71, 77, 100, 136
Minimum principle, 2, 3
Minimum-time problem, 3, 72, 85, 113, 139, 160, 167
Mixed-integer programming, 15, 167
Monte Carlo, 198

N

Newton-Raphson method, generalized, 53, 129
Noise, 200
Nonlinear programing, 14, 41, 121, 173, 178, 185, 199, 208, 210
Nonlinear programming, complex variables, 185
Nonlinear systems, continuous-time, 111, 120

Nonlinear systems, discrete-time, 172, 193
Nuclear reactor, 1, 103, 111, 153
Nuclear reactor control, 111
Nuclear reactor fuel recycle, 153
Nuclear reactor, rocket, 103
Nuclear reactor, shutdown, 113

O

Objective function, 9, 12, 14, 25, 30, 38, 41
Optimal control, 1, 2, 4, 25, 59, 62, 70, 111, 172
Optimal control, stochastic systems, 197
Optimal solution, 9, 12, 14
Optimal trajectory, 3, 4

P

Parametric programming problem, 96
Penalty functions, 50
Penalty method, 117
Penalty method, interior, 131
Performance index, 2, 4, 9, 61, 71, 114, 117
Pivot, 33
Primal problem, 21, 91
Principle of optimality, 4
Projection algorithm, 54
Projection matrix, 47

Q

Quadratic programming, 14, 37, 70, 136, 138, 141, 169, 202, 221
Quantized control, 158

R

Rendezvous problem, 79
Rocket problem, 118

S

Saddle point, 19
Sampled data systems, 141, 158, 197
Sampled data systems, distributed-parameter, 222
Sampled data systems, nonlinear, 182, 187
Sampled data systems, statistical design, 200
Sampling, 172
Search method, cubic interpolation, 118
Search method, Fibonacci, modified, 118
Search method, Newton, modified, 118
Simplex algorithm, 25, 27, 40, 93
Simpson's formula, 221
Slack variables, 25, 26, 139
Stabilization, 182
State equations, 1, 2, 71
State variables, 2
Steepest ascent, 43, 52
Stochastic programming, 15
Stochastic systems, 197
SUMT—sequential unconstrained minimization technique, 50, 77, 117, 124, 178, 189, 191

T

Tableau, simplex, 29, 31, 35
Terminal-control problem, 3, 112, 131
Time-optimal problem, 61, 85, 139
Tracking system, 178, 182
Trajectory optimization, 117
Transversality condition, 4
Two-point boundary-value problem, 3, 6, 132

U

Unconstrained minimization, 50, 52, 118, 129, 131, 185
Unconstrained problem, 9, 50

V

Variable metric methods, 52

W

Wolfe's method in QP, 39

X

Xenon poisoning in nuclear reactors, 111